Batten Disease:
Diagnosis, Treatment, and Research

Advances in Genetics

Serial Editors

Jeffery C. Hall
Waltham, Massachusetts

Jay C. Dunlap
Hanover, New Hampshire

Theodore Friedmann
La Jolla, California

Francesco Giannelli
London, United Kingdom

Batten Disease:
Diagnosis, Treatment, and Research

Edited by
Krystyna E. Wisniewski
Nanbert Zhong
New York State Institute for Basic Research
in Developmental Disabilities
Staten Island, New York

ACADEMIC PRESS
A Harcourt Science and Technology Company

San Diego   San Francisco   New York
Boston   London   Sydney   Tokyo

This book is printed on acid-free paper.

Copyright © 2001 by ACADEMIC PRESS

All Rights Reserved.
No part of this publication may be reproduced or transmitted in any form or by any means, electronic or mechanical, including photocopy, recording, or any information storage and retrieval system, without permission in writing from the Publisher.

The appearance of the code at the bottom of the first page of a chapter in this book indicates the Publisher's consent that copies of the chapter may be made for personal or internal use of specific clients. This consent is given on the condition, however, that the copier pay the stated per copy fee through the Copyright Clearance Center, Inc. (222 Rosewood Drive, Danvers, Massachusetts 01923), for copying beyond that permitted by Sections 107 or 108 of the U.S. Copyright Law. This consent does not extend to other kinds of copying, such as copying for general distribution, for advertising or promotional purposes, for creating new collective works, or for resale. Copy fees for pre-2001 chapters are as shown on the title pages. If no fee code appears on the title page, the copy fee is the same as for current chapters.
0065-2660/01 $35.00

Explicit permission from Academic Press is not required to reproduce a maximum of two figures or tables from an Academic Press chapter in another scientific or research publication provided that the material has not been credited to another source and that full credit to the Academic Press chapter is given.

Academic Press
A Harcourt Science and Technology Company
525 B Street, Suite 1900, San Diego, California 92101-4495, USA
http://www.academicpress.com

Academic Press
Harcourt Place, 32 Jamestown Road, London NW1 7BY, UK
http://www.academicpress.com

International Standard Book Number: 0-12-017645-9

PRINTED IN THE UNITED STATES OF AMERICA
01 02 03 04 05 06 EB 9 8 7 6 5 4 3 2 1

Contents

Contributors   xi
Preface   xiii

1 Neuronal Ceroid Lipofuscinoses: Classification and Diagnosis   1
Krystyna E. Wisniewski, Elizabeth Kida, Adam A. Golabek, Wojciech Kaczmarski, Fred Connell, and Nan Zhong

    I. Introduction   3
    II. Current Classification of NCLs   10
    III. CLN1. Diagnostic Criteria and Phenotype–Genotype Correlation   12
    IV. CLN2. Classic Late-Infantile NCL   16
    V. CLN3. Juvenile NCL   19
    VI. CLN4. Adult NCL   24
    VII. CLN5. Finnish Late-Infantile Variant   24
    VIII. CLN6. Variant Late-Infantile Gypsy/Indian   25
    IX. CLN7. Turkish Variant Late-Infantile NCL   25
    X. CLN8. Northern Epilepsy   29
    XI. Summary   29
    References   30

2 Cellular Pathology and Pathogenic Aspects of Neuronal Ceroid Lipofuscinoses   35
Elizabeth Kida, Adam A. Golabek, and Krystyna E. Wisniewski

    I. Introduction   36
    II. CLN1. Infantile Form of NCL: Deficiency of Palmitoyl-Protein Thioesterase 1   37
    III. CLN2. The Classic Late-Infantile NCL: Deficiency of Tripeptidyl-Peptidase I   42
    IV. CLN3. Juvenile Form of NCL: Genetic Defect of Lysosomal Membrane Protein   49

V. Other NCL Forms with Known Genetic Defect   54
VI. NCL Forms without Identified Genetic Defects   56
  References   59

3 Positional Candidate Gene Cloning of CLN1   69
Sandra L. Hofmann, Amit K. Das, Jui-Yun Lu, and Abigail A. Soyombo

I. Introduction   70
II. Linkage Disequilibrium Mapping of the *CLN1* Locus in the Finnish Population   71
III. Palmitoyl-Protein Thioesterase Defines a New Pathway in Lysosomal Catabolism   72
IV. Enzymology of PPT   73
V. Posttranslational Processing and Lysosomal Targeting of PPT   75
VI. The Physiological Role of PPT   77
VII. Palmitoyl-Protein Thioesterase-2 (PPT2)   79
VIII. The PPT cDNA and Gene   79
IX. The Molecular Genetics of CLN1/PPT Deficiency   81
X. Laboratory Diagnosis of PPT Deficiency   86
XI. Prospects for Cause-Specific Treatment of PPT Deficiency   86
  References   88

4 Biochemistry of Neuronal Ceroid Lipofuscinoses   93
Mohammed A. Junaid and Raju K. Pullarkat

I. Introduction   94
II. Genetic Defects   95
III. NCL are Lysosomal Storage Diseases   99
IV. Remaining Issues and Future Directions   101
  References   103

5 Positional Cloning of the JNCL Gene, *CLN3*   107
Terry J. Lerner

I. Introduction   107
II. *CLN3* Maps to Chromosome 16   108
III. A Subunit 9 Gene Is Not *CLN3*   109
IV. Refined Localization of *CLN3*   109

V. Physical Mapping of the CLN3 Candidate Region   111
VI. Exon Trapping Yields a Candidate cDNA   113
VII. The Common Mutation in JNCL is a 1-kb Genomic Deletion   113
VIII. Mutational Analysis of the CLN3 Gene   115
IX. Tissue Expression of CLN3   116
X. CLN3 Encodes a Novel Protein   116
XI. Animal Models of JNCL   117
References   118

6 Studies of Homogenous Populations: CLN5 and CLN8   123

Susanna Ranta, Minna Savukoski, Pirkko Santavuori, and Matti Haltia

I. Introduction   125
II. Clinical Data   125
III. Neurophysiology   127
IV. Neuroradiology   128
V. Morphology, Cytochemistry, and Biochemistry   128
VI. Molecular Genetics and Cell Biology   132
VII. Diagnosis   135
VIII. Treatment   137
IX. Mouse Homolog for CLN8   137
References   138

7 Molecular Genetic Testing for Neuronal Ceroid Lipofuscinoses   141

Nanbert Zhong

I. Introduction   142
II. Specimens Required for Genetic Testing   143
III. Molecular Genetic Testing for JNCL   144
IV. Molecular Genetic Testing for LINCL and INCL   149
V. Molecular Screening of Carrier Status in NCL Families   151
VI. Prenatal Diagnostic Testing for NCL   152
VII. Important Issues in the Molecular Genetic Testing   154
References   156

8 Genetic Counseling in the Neuronal Ceroid Lipofuscinoses 159
Susan Sklower Brooks

    I. Introduction   159
    II. Inheritance   160
    III. Genetics   160
    IV. Diagnostic Confirmation   161
    V. Genetic Counseling   161
    VI. Carrier Screening   162
    VII. Reproductive Options   164
    VIII. Conclusion   165
    References   166

9 Neurotrophic Factors as Potential Therapeutic Agents in Neuronal Ceroid Lipofuscinosis 169
Jonathan D. Cooper and William C. Mobley

    I. Introduction   170
    II. Mouse Models of NCLs   170
    III. Characterization of the CNS of Mouse Models of NCLs   171
    IV. Neurotrophic Factors as Potential Therapeutic Agents in Neurodegenerative Disorders?—The Neurotrophic Factor Hypothesis   173
    V. NTF Expression and Actions beyond the "Neurotrophic Factor Hypothesis"   176
    VI. Failure of NTF Signaling—A Cause of Neuronal Dysfunction and Degeneration?   177
    VII. Treatment with IGF-1—Implications for the Treatment of NCLs   178
    VIII. Toward Clinical Trials of NTFs   179
    References   179

10 Animal Models for the Ceroid Lipofuscinoses 183
Martin L. Katz, Hisashi Shibuya, and Gary S. Johnson

    I. The Need for Animal Models   184
    II. The Human Disorders   184
    III. Naturally Occurring Ceroid Lipofuscinosis in Animals as Models for the Human Disorders   186

IV. Animal Models Created through Molecular
Genetic Manipulation   195
V. Future Directions   198
References   199

11 Experimental Models of NCL: The Yeast Model   205
David A. Pearce

I. Introduction   205
II. Yeast as a Model for JNCL   206
III. What Does Btn1p Do?   210
IV. Yeast as a Therapeutic Model for JNCL   213
V. A Yeast Model for INCL   214
References   215

12 Outlook for Future Treatment   217
Nanbert Zhong and Krystyna E. Wisniewski

I. Molecular Cloning for CLN_4, CLN_6, and CLN_7 218
II. Characterization of Native Substrates for CLN-Encoded Lysosomal Enzymes   218
III. Proteomic Studies of CLN-Encoded Proteins   219
IV. Uncovering the Pathogenesis of the NCLs   219
V. Potential Drugs in Experimental Models May Eventually Lead to Clinical Trials in NCL-Affected Patients   220
VI. Gene Therapy   221
References   222

Appendix: Batten Support Groups   225

United States of America   225
Canadian Chapter   230
European Support Groups   230
Elsewhere   235

Index   237

Contributors

Numbers in parentheses indicate the pages on which the authors' contributions begin.

Susan Sklower Brooks (159) Department of Human Genetics, New York State Institute for Basic Research in Developmental Disabilities, Staten Island, New York 10314

Fred Connell (1) Department of Pathological Neurobiology, New York State Institute for Basic Research in Developmental Disabilities, Staten Island, New York 10314

Jonathan D. Cooper (169) Department of Neurology and Neurological Sciences, and the Program in Neuroscience, Stanford University School of Medicine, Stanford, California 94305

Amit K. Das (69) Department of Internal Medicine and the Hamon Center for Therapeutic Oncology Research, University of Texas Southwestern Medical Center, Dallas, Texas 75390

Adam A. Golabek (1, 35) Department of Pathological Neurobiology, New York State Institute for Basic Research in Developmental Disabilities, Staten Island, New York 10314

Matti Haltia (123) Department of Pathology, University of Helsinki and Helsinki University Central Hospital, 00014 Helsinki, Finland

Sandra L. Hofmann (69) Department of Internal Medicine and the Hamon Center for Therapeutic Oncology Research, University of Texas Southwestern Medical Center, Dallas, Texas 75390

Gary S. Johnson (183) Department of Veterinary Pathobiology, University of Missouri College of Veterinary Medicine, Columbia, Missouri 65211

Mohammed A. Junaid (93) Department of Developmental Biochemistry, New York State Institute for Basic Research in Developmental Disabilities, Staten Island, New York 10314

Wojciech Kaczmarski (1) Department of Pathological Neurobiology, New York State Institute for Basic Research in Developmental Disabilities, Staten Island, New York 10314

Martin L. Katz (183) University of Missouri School of Medicine, Mason Eye Institute, Columbia, Missouri 65212

Elizabeth Kida (1, 35) Department of Pathological Neurobiology, New York State

Institute for Basic Research in Developmental Disabilities, Staten Island, New York 10314

Terry J. Lerner (107) Molecular Neurogenetics Unit, Massachusetts General Hospital, Charlestown, Massachusetts 02129, and Department of Neurology, Harvard Medical School, Boston, Massachusetts 02114

Jui-Yun Lu (69) Department of Internal Medicine and the Hamon Center for Therapeutic Oncology Research, University of Texas Southwestern Medical Center, Dallas, Texas 75390

William C. Mobley (169) Department of Neurology and Neurological Sciences, and the Program in Neuroscience, Stanford University School of Medicine, Stanford, California 94305

David A. Pearce (205) Center for Aging and Developmental Biology, Department of Biochemistry and Biophysics, University of Rochester School of Medicine and Dentistry, Rochester, New York 14642

Raju K. Pullarkat (93) Department of Developmental Biochemistry, New York State Institute for Basic Research in Developmental Disabilities, Staten Island, New York 10314

Susanna Ranta (123) Department of Molecular Genetics, The Folkhälsan Institute of Genetics, 00280 Helsinki, Finland, and Department of Medical Genetics, University of Helsinki, 00014 Helsinki, Finland

Pirkko Santavuori (123) Department of Neurology, Hospital for Children and Adolescents, Helsinki University Central Hospital, 00029 Helsinki, Finland

Minna Savukoski (123) Department of Human Molecular Genetics, National Public Health Institute, 00300 Helsinki, Finland, and Department of Medical Genetics, University of Helsinki, 00014 Helsinki, Finland

Hisashi Shibuya (183) Department of Veterinary Medicine, Nihon University, Fujisawa, Japan 252-8510

Abigail A. Soyombo (69) Department of Internal Medicine and the Hamon Center for Therapeutic Oncology Research, University of Texas Southwestern Medical Center, Dallas, Texas 75390

Krystyna E. Wisniewski (1, 35, 217) Department of Pathological Neurobiology, New York State Institute for Basic Research in Developmental Disabilities, Staten Island, New York 10314, and Department of Pediatric Neurology, State University of New York/Health Science Center, Brooklyn, New York 11203

Nanbert Zhong (1, 141, 217) Department of Human Genetics, New York State Institute for Basic Research in Developmental Disabilities, Staten Island, New York 10314, and Department of Pediatric Neurobiology, State University of New York/Health Science Center, Brooklyn, New York 11203

Preface

The highest goal of science, as it is applied to medicine, is to alleviate human suffering. Current research on genes and the human brain focuses on diseases and disorders of the central nervous system, with a broader perspective of obtaining an understanding of how the brain works and how it can be helped to work better if there is a problem.

The neuronal ceroid lipofuscinoses (NCLs) are a group of inherited disorders that cause progressive neurological dysfunction. The clinical characteristics may vary from individual to individual and include retinal degeneration, which can lead to progressive blindness, cognitive and/or motor dysfunction, and epilepsy. These disorders mainly affect infants and children; rarely adults. In the past, the NCLs were diagnosed and categorized into four forms on the basis of age-at-onset and ultrastructural findings. Recent progress in biochemical and molecular genetics inclined us to reevaluate some atypical NCL cases. Eight forms have now been characterized, with numerous common and uncommon mutations, and more are to be identified. Prevention through early diagnosis, carrier detection, and genetic counseling is now available. Neuroscience, like other disciplines, set out to identify the cellular and molecular building blocks, as well as organizational principles and processes of the central nervous system. As multidisciplinary approaches address the causes of NCL, the underlying defect is still unknown. Hopefully, within the next decade, we will gain a better understanding of the defect. To date, no effective therapy for NCL is available; therefore, more research is vital. These twelve chapters describe the current status of NCL diagnosis, treatment, and research.

Krystyna E. Wisniewski
Nanbert Zhong

1
Neuronal Ceroid Lipofuscinoses: Classification and Diagnosis

Krystyna E. Wisniewski*
Department of Pathological Neurobiology
New York State Institute for Basic Research in Developmental Disabilities
Staten Island, New York 10314
and
Department of Pediatric Neurology
State University of New York/Health Science Center
Brooklyn, New York 11203

Elizabeth Kida, Adam A. Golabek, Wojciech Kaczmarski, and Fred Connell
Department of Pathological Neurobiology
New York State Institute for Basic Research in Developmental Disabilities
Staten Island, New York 10314

Nanbert Zhong
Department of Human Genetics
New York State Institute for Basic Research in Developmental Disabilities
Staten Island, New York 10314
and
Department of Pediatric Neurobiology
State University of New York/Health Science Center
Brooklyn, New York 11203

    I. Introduction
   II. Current Classification of NCLs
  III. CLN1. Diagnostic Criteria and Phenotype–Genotype Correlation
       A. Classic Infantile Onset with GROD

*Address for correspondence: New York State Institute for Basic Research in Developmental Disabilities, 1050 Forest Hill Road, Staten Island, New York 10314. E-mail: BATTENKW@AOL.COM

       B. Variant Late-Infantile Onset with
          GROD
       C. Variant Juvenile Onset with GROD
  IV. CLN2. Classic Late-Infantile NCL
   V. CLN3. Juvenile NCL
  VI. CLN4. Adult NCL
 VII. CLN5. Finnish Late-Infantile Variant
VIII. CLN6. Variant Late-Infantile Gypsy/Indian
  IX. CLN7. Turkish Variant Late-Infantile NCL
   X. CLN8. Northern Epilepsy
  XI. Summary
       References

ABSTRACT

The neuronal ceroid lipofuscinoses (NCLs) are neurodegenerative disorders characterized by accumulation of ceroid lipopigment in lysosomes in various tissues and organs. The childhood forms of the NCLs represent the most common neurogenetic disorders of childhood and are inherited in an autosomal-recessive mode. The adult form of NCL is rare and shows either an autosomal-recessive or autosomal dominant mode of inheritance. Currently, five genes associated with various childhood forms of NCLs, designated CLN1, CLN2, CLN3, CLN5, and CLN8, have been isolated and characterized. Two of these genes, CLN1 and CLN2, encode lysosomal enzymes: palmitoyl protein thioesterase 1 (PPT1) and tripetidyl peptidase 1 (TPP1), respectively. CLN3, CLN5, and CLN8 encode proteins of predicted transmembrane topology, whose function has not been characterized yet. Two other genes, CLN6 and CLN7, have been assigned recently to small chromosomal regions. Gene(s) associated with the adult form of NCLs (CLN4) are at present unknown. This study summarizes the current classification and new diagnostic criteria of NCLs based on clinicopathological, biochemical, and molecular genetic data. Material includes 159 probands with NCL (37 CLNI, 72 classical CLN2, 10 variant LINCL, and 40 CLN3) collected at the New York State Institute for Basic Research in Developmental Disabilities (IBR) as well as a comprehensive review of the literature. The results of our study indicate that although only biochemical and molecular genetic studies allow for definitive diagnosis, ultrastructural studies of the biopsy material are still very useful. Thus, although treatments for NCLs are not available at present, the diagnosis has become better defined.

I. INTRODUCTION

The neuronal ceroid lipofuscinoses (NCLs, Batten disease), are common, progressive neurological disorders in infancy and childhood, with an incidence ranging in various countries from 0.1 to 7 per 100,000 live births (Santavuori, 1988; Uvebrant and Hagberg, 1997). Childhood forms of NCLs are inherited in an autosomal-recessive mode. A rare adult form of NCL shows both autosomal-recessive and dominant modes of inheritance. Accumulation of autofluorescent ceroid lipopigment in lysosomes of various tissues and organs is a characteristic feature of NCLs.

The term neuronal ceroid lipofuscinosis was introduced by Zeman and Dyken (1969) to clearly differentiate this group of disorders from the gangliosidoses. However, as far back as the Middle Ages, or perhaps even earlier, people were suffering from occulo-cerebral dysfunction manifested by progressive blindness, epilepsy, and dementia, although the nature of these disorders remained unknown. The first description of this disease, which we could classify currently as a juvenile variant of NCL, comes from Stengel, who in 1826 presented four siblings demonstrating progressive blindness, epilepsy, cognitive decline, and motor dysfunction. Other NCL cases with juvenile onset were described by Batten (1903), Spielmeyer (1923), and Hallervorden (1938). Janský (1908) and Bielschowsky (1923) documented NCL subjects with late-infantile onset, and Kufs (1925) described the adult variant of the disease. Haltia et al. first described the infantile form of NCL (1973).

Further electron microscopical (EM) studies of various biopsy and autopsy tissues revealed that lysosomal storage of NCL subjects may contain four types of inclusion bodies: granular osmiophilic deposits (GROD), curvilinear profiles (CV), fingerprint profiles (FP), and rectilinear complexes (RL) (Santavuori, 1988; Wisniewski et al., 1988, 1992, 1998a, 1998b, 1999, 2000; Goebel et al., 1999; and Elleder et al., 1999). Thus, on the basis of age at onset, clinical course, and pathological findings, NCLs were traditionally divided into four major forms. The infantile form (INCL, Santavuori-Haltia disease) had onset of symptoms between 6 months and 2 years and was characterized by progressive dementia (intellectual decline), motor dysfunction, blindness, and seizures, and the presence of GROD evidenced by EM (Figure 1.1).

The late-infantile form (LINCL, Janský-Bielschowsky disease) had onset of symptoms between 2 and 4 years of age, with progressive dementia, blindness, motor dysfunction, and seizures, and CV profiles evident with EM (Figure 1.2).

The juvenile form (JNCL, Batten-Spielmeyer-Vogt disease) had onset of symptoms between 4 and 10 years of age, with progressive blindness, cognitive and motor dysfunction, and seizures, and FP and vacuolated lymphocytes evident with EM (Figures 1.3 and 1.4).

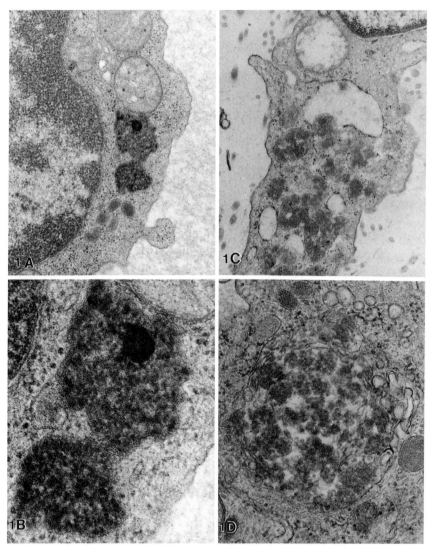

Figure 1.1. (A) CLN1 infantile neuronal ceroid lipofuscinosis (NCL). Two densely packed lipopigments of granular osmiophilic deposits (GROD) in lysosomes, localized in the cytoplasm of lymphocyte (buffy coat), × 42,500. (B) Higher magnification of (A), × 241,000. (C) Numerous densely and loosely packed GROD localized in the lysosomal storage material in fibroblast from a skin-punch biopsy, × 39,000. (D) Higher magnification of (C), × 98,500.

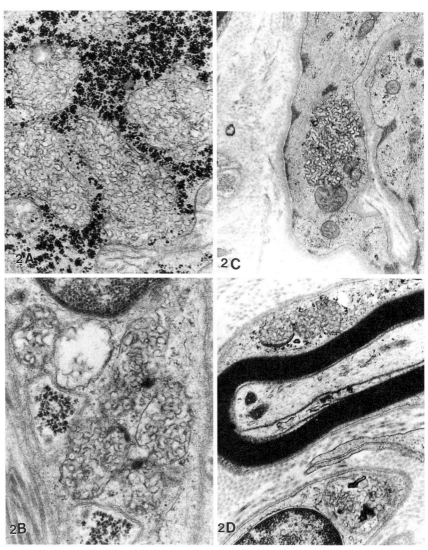

Figure 1.2. (A) CLN2 late-infantile NCL. The secretory exocrine sweat glands with numerous lysosomes storing material composed of curvilinear profiles, × 53,000. (B) Fibroblast with numerous lysosomes that are storing fingerprint profiles, × 79,000. (C) In a pericyte of vessel wall, lysosomal storage with curvilinear profiles, × 23,500. (D) Subcutaneous nerve bundles showing, in the cytoplasm of Schwann cells, lysosomes that are storing curvilinear profiles, × 31,000.

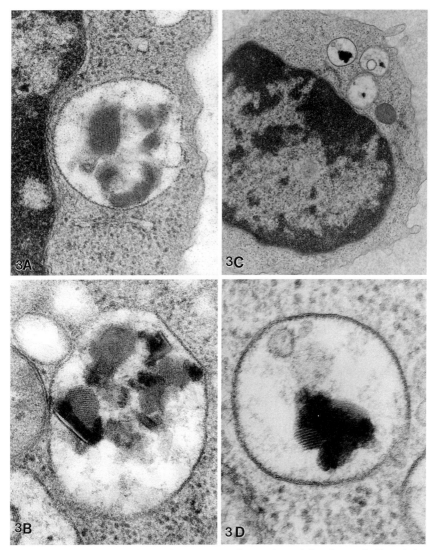

Figure 1.3. (A) CLN3, juvenile NCL, homozygous. In the cytoplasm of a lymphocyte (buffy coat), fingerprint profiles within membrane-bound lysosomal vacuoles, × 64,000. (B) Higher magnification of (A), × 73,500. (C) Lymphocytes with fingerprint profiles within membrane-bound lysosomal vacuoles, similar to (A) and (B), × 27,000. (D) Higher magnification of (C), × 225,000.

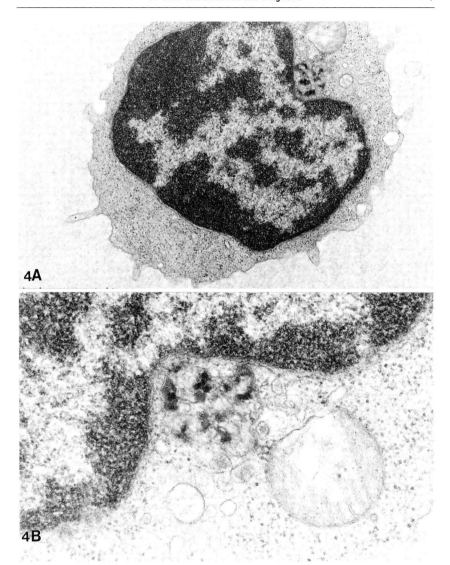

Figure 1.4. (A) CLN3, juvenile NCL, compound heterozygous. In the cytoplast of lymphocytes (buffy coat) are large lysosomes that are storing mixed (curvilinear and fingerprint) profiles, × 30,180. (B) Higher magnification of (A), × 69,800.

Table 1.1. NCL—Batten Disease Registry at IBR

Form	No.	%	Within USA	%	Outside USA	%	Age of death (yr) Range	Mean
INCL	107	11.8	83	10.8	24	16.8	1.42–26	9.37
LINCL	381	41.9	305	39.8	76	53.1	4.25–43.25	9.84
JNCL	402	44.2	359	46.9	43	30.1	7.92–51.17	20.81
Kufs	19	2.1	19	2.5	0	0	32.08–64	36.77
Totals	909	100	766	100	143	100		

The adult form (ANCL, Kufs disease) had onset after 15 years of age, either with myoclonic seizures or progressive behavioral, cognitive, and motor dysfunction, and mixed lysosomal inclusions (CV, FP, GROD) by EM (Figure 1.5)

The distribution of major NCL forms collected through the Batten Disease Registry at the New York State Institute for Basic Research in Developmental Disabilities (IBR) is presented in Table 1.1. The juvenile form is the most common form in the United States. The second most common is the late-infantile form. With the advent of molecular testing for INCL, the number of infantile forms is dramatically increasing. This form is not exclusive to the Finnish population, as originally thought, but may exist in all populations.

However, in the 1970s and 1980s, clinicopathological correlations showed that around 20% of NCL cases did not fit into these four classical forms (Wisniewski et al., 1988, 1992; Dyken and Wisniewski 1995; Wisniewski et al., 1999). Further, rapid progress in molecular genetic studies of NCLs was possible as a result of the formation of the International Batten Disease Consortium. The combined efforts of clinicians, pathologists, and molecular geneticists has allowed identification of new forms of NCL, isolation and characterization of five genes associated with the NCL disease process, and mapping of two other genes. In this chapter, a new classification of NCLs that has emerged from these studies as well as new diagnostic criteria based on clinicopathological, biochemical, and molecular genetic data will be summarized and discussed. This study is based on analysis of 159 probands with NCL collected at the George Jervis Clinic at IBR, including data obtained from the Batten Disease Registry at IBR. The ultrastructural studies were performed similarly to those discussed previously (Wisniewski et al., 1988, 1992, 1998a, 1998b). The biochemical and molecular genetic studies were carried out as described elsewhere (Das et al., 1998; Wisniewski et al., 1998a, 1999; Zhong et al., 1998a, 1998b, 2000; Sleat et al., 1999; Hartikainen et al., 1999; Junaid et al., 1999; Mole et al., 1999; Hofmann et al., 1999; see also Chapters 3–5). A comprehensive summary of data obtained by other researchers also will be included.

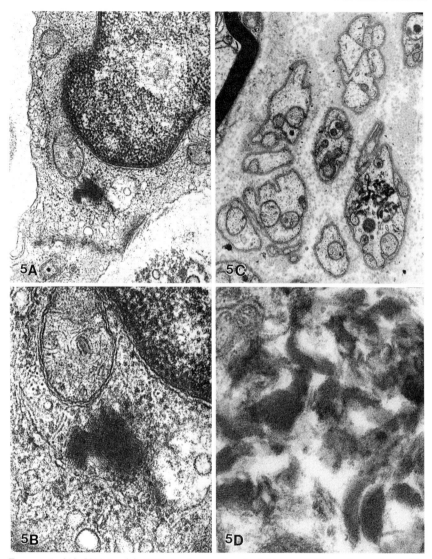

Figure 1.5. (A) CLN 4, adult NCL. In the vessel wall of skin punch biopsy, fingerprint profiles without membrane-bound lysosomal vacuoles, × 32,500. (B) Higher magnification of (A), × 82,300. (C) Unmyelinated nerve fibers with fingerprint profiles, × 21,000. (D) Numerous fingerprint and trilamellar structures in the lysosomal storage material, × 220,000.

II. CURRENT CLASSIFICATION OF NCLs

The diagnosis of NCL is based at present on a careful analysis of data provided by clinicopathological, biochemical, and genetic studies. Currently, eight forms of NCLs can be distinguished, designated CLN1–CLN8. Five genes associated with the NCL disease process have been isolated and characterized: CLN1, CLN2, CLN3, CLN5, and CLN8 (Vesa et al., 1995; IBDC 1995; Sleat et al., 1997; Savukoski et al., 1998; Ranta et al., 1999). Two genes, CLN6 and CLN7, have been mapped to a small chromosome region (Sharp et al., 1997; Williams et al., 1999). Isolation of gene(s) associated with the adult variant of NCLs (CLN4) is still in progress.

The functions of two products of NCL-associated genes, CLN1 and CLN2, are known. CLN1 gene encodes palmitoyl protein thioesterase (PPT1), a lysosomal enzyme involved in depalmitoylation of proteins (Camp and Hofmann 1993; Camp et al., 1994). The CLN2 gene encodes a tripeptidyl peptidase 1 (TPP 1), a lysosomal enzyme acting as an aminopeptidase that removes tripeptides from the free N-termini of proteins (Rawlings and Barrett, 1999; Vines and Warburton, 1999). The functions of three other proteins encoded by CLN3, CLN5, and CLN8 genes are unknown at present; however, their predicted amino acid sequences suggest that these proteins have transmembrane topology (Janes et al., 1996; Savukoski et al., 1998; Ranta et al., 1999; see also Chapter 6)

To date, over 80 different mutations and polymorphisms have been identified in NCL-associated genes, and more mutations are expected to be found (Munroe et al., 1997; Hofmann et al., 1999; Sleat et al., 1999; Mole et al., 1999; Wisniewski et al., 2000; see also Chapter 3): about 30 each of CLN1 and CLN2, 25 CLN3 and 5 of CLN5 (Figures 1.6–1.8).

These mutations can be divided into two groups: (1) common or major mutations that predominate in a given form of NCL and are associated with the classical course of the disease, and (2) rare, private mutations that occur sporadically and may be associated with a different phenotype than the classical one. As was revealed by phenotype–genotype correlation, the age at onset of the disease process can no longer represent a criterion for the assignment of a case as a particular NCL form. It became evident that mutations in the CLN1 gene can be associated with infantile, late-infantile, and juvenile onset of the disease process. Mutations in the CLN2 gene may give rise not only to late-infantile onset of the disease, but also to juvenile onset. Late-infantile onset of the disease may be associated with mutations in one of any five genes: CLN1, CLN2, CLN5, CLN6, or CLN7. Furthermore, the results of ultrastructural studies of biopsy material also may be misleading, considering that mixed types of lysosomal inclusion bodies may be found (Wisniewski et al., 1988, 1992). All these findings indicate that only a

CLN1 Mutations (Total=34)

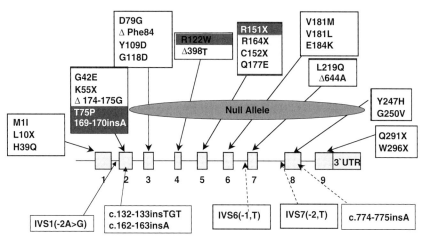

Figure 1.6. CLN1 mutations.

CLN2 Mutations (Total: 32)

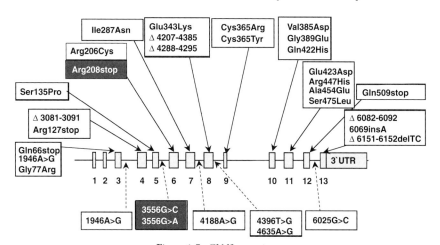

Figure 1.7. CLN2 mutations.

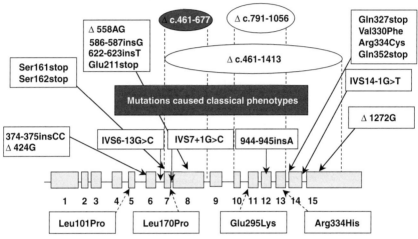

Figure 1.8. CLN3 mutations.

careful evaluation of all data provided by clinicopathological, biochemical, and genetic studies can guarantee an accurate diagnosis. The proper classification of a particular form of NCL is no longer a purely nosological problem because, first, therapeutic approaches of NCL have already been initiated (Lake et al., 1997; Vanhanen et al., 1998; see also Chapters 7 and 12). The criteria used currently for diagnosis of particular forms of NCL are summarized in Table 1.2.

III. CLN1. DIAGNOSTIC CRITERIA AND PHENOTYPE–GENOTYPE CORRELATION

Mutations in the PPT 1 gene can give rise to three major phenotypes: ones with the classic infantile onset, the variant late-infantile onset, and the variant juvenile onset. Of 37 cases evaluated at IBR and assigned to CLN1, the classic infantile-onset form was found in 18, the variant late infantile onset with GROD in 7, and the variant juvenile onset with GROD in 12. Thus, only around 50% of cases demonstrated infantile onset of the disease process, which in the past was considered typical for this form of NCL (Santavuori et al., 1973, 1989; Santavuori 1988). However, the classic infantile form is the only form of CLN1 encountered

Table 1.2. Classification of Neuronal Ceroid Lipofuscinoses on the Basis of Clinicopathological, Biochemical, and Genetic Testing

Form	Genetic symbol	Onset (age range)	Presenting symptoms	EM results[c]	Mutated protein	Gene location	Common mutation/location	Total # of mutations
Infantile NCL	CLN1[a]	0–37 yr	Visual loss, Motor dysfunction, Intellectual decline	GROD[c]	PPT1[d]	1p32	Arg122Trp Exon 4, Arg151STOP Exon 5, Thr75Pro Exon 2	>30
Classic late infantile NCL	CLN2[b]	2–8 yr	Seizures, Motor dysfunction, Intellectual decline	CV/mixed	TPP 1	11p15	IVS5-1G C Intron 5, Arg208STOP Exon 6	>30
Juvenile NCL	CLN3	4–10 yr	Visual loss, Intellectual, motor decline, seizures	FP, or mixed, vacuolated lymphocytes	Battenin Transmembrane	16p12	1.02-kb deletion Intron 6-8	>20
Adult NCL	CLN4	11–55 yr	A: myoclonus epilepsy with dementia; B: behavior and dementia	Mixed	Unknown	Unknown	Unknown	Unknown
Finnish LINCL	CLN5	4–7 yr	Motor dysfunction, Visual loss, Intellectual decline	FP, CV, RL	Transmembrane	13q22	Tyr392STOP Exon 4	3 mutations, 1 polymorphism
Gypsy/Indian LINCL	CLN6	18 mo–8 yr	Motor dysfunction, Intellectual decline, Seizures, Visual loss	CV, FP, RL	Unknown	15q21-q23	Unknown	Unknown
Turkish LINCL	CLN7	1–6 yr	Seizures, Motor dysfunction, Visual loss	FP or mixed	Unknown	Unknown	Unknown	Unknown
EPMR[d]	CLN8	1–10 yr	Seizures, Intellectual decline	CV- or GROD-like structures	Transmembrane	8p23	Arg24Gly Exon 2	1

[a]Onset: infantile, 6 mo–2 yr; late infantile, 2–4 yr; juvenile, 5–10 yr; adult, (>10 yr) in all forms of NCL.
[b]Onset: late infantile, 2–4 yr; juvenile, 4–7 yr.
[c]EM, electron microscopy; CV, curvilinear profiles; FP, fingerprint profiles; GROD, granular osmiophilic deposits; RL, rectilinear complex. Mixed: CV, FP, RL, GROD. PPT1, palmitoyl-protein thioesterase 1; TPP 1, tripeptidyl peptidase.
[d]EPMR: progressive epilepsy with mental retardation.

in Finland. Finland also has the highest incidence of mutated CLN1 gene—1 per 20,000 live births—and the highest carrier frequency: 1 in 70.

A. Classic infantile onset with GROD

In the majority of subjects with the classic infantile form of NCL (INCL), the first clinical signs and symptoms appear at the age of 6–18 months. Early symptoms and signs include retarded growth of the head, muscular hypotonia, excitability, and lack of motor-skill development. Progressive loss of vision (starting around 15–22 months), intellectual decline (starting around 10–18 months), ataxia (starting around 12–24 months), myoclonus (starting around 16–36 months), involuntary movements (starting around 7–20 months), and epileptic seizures (starting around 15 months–3.6 years), mostly of simple or complex partial type, complete the clinical picture. Brownish discoloration of the macula, retinal degeneration, and optic atrophy are present after 2 years of age, but pigment aggregation, typical for CLN2 and CLN3 subjects, is usually absent. In the final stage of the disease, the affected children are blind and demonstrate spasticity and myoclonia; however, some emotional contact can be maintained with them even until death. Children die usually at the age of 8–13 years (Santavuori et al., 1999).

Electrophysiological studies may be helpful to establish the final diagnosis. Electroretinograms (ERG) are abnormal at around the age of 2.5 years. Visual evoked potentials (VEP) are abolished around the age of 2–5 years. Abnormalities of electroencephalograms (EEG) appear around the age of 2 years, and EEG often becomes flat around the age of 3 years. Magnetic resonance images (MRI) show abnormalities early, around the age of 7–10 months. T-weighted images disclose hypointensity of the thalami to the white matter and basal ganglia, and later on, high signal intensity of the white matter and severe progressive cerebral and cerebellar atrophy.

EM studies of diagnostic biopsy specimens (e.g., conjunctiva, skin, and/or rectal biopsies, and/or lymphocytes [buffy coat]) are recommended. GROD have been found very early during the disease process (eighth week of pregnancy); analysis of chorionic villi from "at risk" pregnancies has been used successfully in the past for prenatal diagnosis (Rapola et al., 1990) and postnatally (Das et al., 1998; Hofmann et al., 1999).

However, definitive diagnosis of INCL is based currently on biochemical and molecular genetic studies. All children with the classic infantile form have either undetected or marginal PPT1 activity (Vesa et al., 1995; Das et al., 1998; Wisniewski et al., 1998b; Waliany et al., 1999; Hofmann et al., 1999; Mole et al., 1999; see also Chapter 3). Recently, a new, sensitive and simple biochemical assay was developed, which allows measurement of PPT1 activity in various biological specimens such as fibroblasts, leukocytes, amniotic fluid cells, chorionic villi, plasma, and cerebrospinal fluid (van Diggelen et al., 1999). Molecular genetic studies identified 24 mutations associated with this particular form of NCLs. In Finland, only one missense mutation (Arg122Trp) was found in all affected children (Vesa et al., 1995; Santavuori et al., 1999). In our material, however, 14 different mutations were associated with classic infantile phenotype, and Arg122Trp was found only in 3/73 CLN1 alleles examined (Das et al., 1998; Waliany et al., 1999). A nonsense point mutation, Arg151Stop, was the most common in the CLN1 probands with classic infantile onset, accounting for 50% of alleles examined in this group. Mutations of CLN1 cases were also identified in Italy (Santorelli et al., 1998).

B. Variant late-infantile onset with GROD

Only seven probands, all from IBR's collection, were assigned to the variant late-infantile onset with GROD (Das et al., 1998; Wisniewski et al., 1998b; Waliany et al., 1999; Hofmann et al., 1999). Clinical signs and symptoms start at around 18 months–3.5 years. Initial symptoms include seizures and intellectual decline; however, in one proband, visual loss also was observed early. Gradually, visual loss and motor and cognitive dysfunction proceed, leading to a vegetative state. Progression of the disease process is slightly slower than in the classic infantile form. Some affected children are still alive at the age of 10–13 years.

Biopsy material analyzed by electron microscopy reveals GROD, but in one proband, an admixture of CV also was disclosed. Electrophysiological studies may be helpful for the final diagnosis; however, definitive diagnosis is based on findings of deficient PPT 1 activity and mutations in the PPT1 gene. Various mutations in the PPT1 gene have been found in the probands analyzed to date, even those that also were described in the classic infantile form (Arg151Stop, Val181Met, Arg164Stop, Y109D). However, three mutations were identified only in children with the late-infantile-onset form: Q177E, V181L, W296X (Das et al., 1998; Wisniewski et al., 1998b; Waliany et al., 1999; see also Chapter 3).

C. Variant juvenile onset with GROD

Numerous NCL subjects with clinical presentation resembling the juvenile NCL form, but demonstrating GROD and no FP in biopsy and autopsy tissues, have been documented (Zeman and Dyken 1969; Carpenter et al., 1973; Manca et al., 1990; Wisniewski et al., 1992; Taratuto et al., 1995; Philippart et al., 1995; Hofman and Taschner 1995; Lake et al., 1996; Crow et al., 1997; Wisniewski et al., 1998b). The nosological status of this group of patients was uncertain. Only recently, molecular genetic and biochemical studies revealed that these patients have mutations in the PPT1 gene and deficient PPT1 activity in various diagnostic specimens, thus representing an allelic variant of CLN1 (Das et al., 1998; Mitchison et al., 1998).

Probands with this variant of CLN1 were reported in various countries, but this variant of NCL occurs at a high frequency in the central region of Scotland (Crow et al., 1997; Stephenson et al., 1999).

This disorder starts at the age of 6–10 years, usually with visual loss. Rarely, intellectual decline or seizures are the initial signs. Hyperactive behavior as the initial sign also was noted (Hofman and Taschner, 1995). Afterwards, motor dysfunction (10–13 years), cognitive dysfunction (11–13 years), and epilepsy with tonic clonic seizures or absences (12–19 years) develop. Subjects usually die after the age of 14 years, but patients still alive after the age of 26 years also have been reported.

The ERG is usually extinguished at the time of diagnosis of retinopathy. The VEP are reduced in amplitude and then become flat. Retinal pigmentary degeneration and macular atrophy are typical; however, lack of ophthalmological changes also was reported (Hofmann and Taschner, 1995). MRI may be normal for a long time. Electron microscopy of the biopsy specimens (conjunctiva, skin, lymphocytes, rectum) shows GROD. In contrast to the juvenile form of NCL (CLN3), in variant juvenile NCL with GROD, lymphocytes show no vacuoles.

Definitive diagnosis is based on detection of PPT1 deficiency and mutations in the PPT1 gene. Of interest, in some probands, residual activity of PPT1 was disclosed, which could at least partially explain the more benign phenotype in these subjects than in children with the classic infantile NCL form (Das et al., 1998). Although numerous mutations were found in this variant of CLN1, most probands are compound heterozygotes for "null"-type mutation and a missense mutation (Das et al., 1998; Mitchison et al., 1998; Mole et al., 1999; see also Chapter 3). The most common mutations are missense point mutations: Thr75Pro, present either in homozygous or heterozygous form; and Arg151Stop, occurring in heterozygous form. Thus, phenotype/genotype comparison shows that mutations in the CLN1 gene can cause three different phenotypes: INCL, LINCL, and JNCL and adult onset (personal communication, Van Diggelen). EM findings do not differ in these various phenotypes; all have GROD (see Figure 1.9). Table 1.3 summarizes the clinicopathological data for CLN1.

IV. CLN2. CLASSIC LATE-INFANTILE NCL

The incidence of classic late-infantile NCL is around 0.36–0.46 per 100,000 live births (Cardona and Rosati, 1995; Claussen et al., 1992). Among the NCL probands evaluated at IBR, 72 subjects were assigned to CLN2. The first symptoms appear at the age of 2–4 years, usually starting with epilepsy, which may be manifested by generalized tonic-clonic seizures, as well as absence, or partial, or secondarily generalized seizures. Regression of developmental milestones is soon

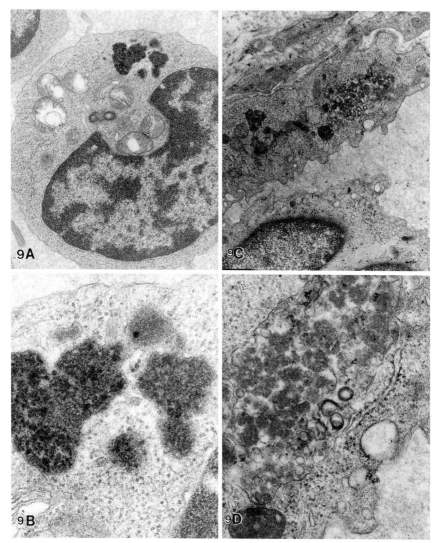

Figure 1.9. (A) CLN1 juvenile (age at onset >4 years). Numerous densely packed GROD in cytoplasm of lymphocyte (buffy coat), × 33,000. (B) Higher magnification of (A), × 130,500. (C) GROD, densely and loosely packed, located in the pericyte of the vessel wall, × 24,000. (D) Higher magnification of (C), × 95,000.

Table 1.3. CLN1 with Infantile, Late-Infantile, and Juvenile Onset

Onset	Clinical signs/findings
Infantile, 6 mo–2 yr	Hypotonia, cognitive dysfunction, seizures, microcephaly, visual failure, ataxia, myoclonic jerks. EEG: disappearance of eye opening/closing reaction, and sleep spindles, low-voltage activity. MRI: severe progressive atrophy
Late infantile, 2–4 yr	Generalized seizures, myoclonic jerks, (may be provoked by flash stimuli during EEG recording), cognitive dysfunction, ataxia, myoclonous, visual failure, motor regression. EEG: occipital photosensitive response using flash rates at 1–2 Hz with eyes open. ERG is diminished. VER enhanced reaching amplitude of 355–375 μV. MRI: progressive brain atrophy
Juvenile, 5–8 yr	Progressive visual failure, cognitive/motor dysfunction (Parkinson-like syndrome), seizures. ERG abolished and VEP abnormal; the later could remain normal for several years. EEG abnormal, frequent paroxysmal activity. MRI: progressive brain atrophy

All CLN1 have GROD at EM level, are PPT1 deficient, and have different types of mutations (see Chapter 3). Onset >6 m and/or >10 yr may also exist in the CLN1 form. Age of death: range 1.42–26 yr.

evident, followed by ataxia and myoclonus. Visual loss leading to blindness appears at the age of 4–6 years. Children are usually bedridden before the age of 6 years.

EEG shows spikes in the occipital region in response to photic stimulation at 1–2 Hz. ERG is usually abnormal at presentation and soon afterwards extinguishes. VEP is enhanced for a long period and diminishes in the final stage of the disease. MRI is helpful for differential diagnosis and shows progressive cerebral and cerebellar atrophy, with normal basal ganglia and thalami. Elevated level of subunit c in urine samples is a constant feature (Wisniewski et al., 1994, 1995). Finding of CV profiles in biopsy material (lymphocytes, skin, conjunctiva, and rectum) facilitates the diagnosis; however, in some cases, mixed lysosomal inclusions (FP, CV, RL, and/or GROD) are observed (Wisniewski et al., 1988, 1992). Ultrastructural studies of amniotic fluid cells, choroid vili and fetal skin biopsy were used for prenatal diagnosis (Chow et al., 1993).

Currently, biochemical assays based on analysis of TPP 1 activity are available to confirm the diagnosis (Sohar et al., 1999; Junaid et al., 1999; Sleat et al., 1999). All CLN2 patients studied to date showed either undetectable levels or severely reduced PPT1 activity in diagnostic specimens (fibroblasts, lymphocytes, brain samples). Molecular genetic studies disclosed 26 mutations and 14 polymorphisms in TPP 1 gene (Sleat et al., 1997; Zhong et al., 1998a; Sleat et al., 1999; Hartikainen et al., 1999; review: Mole et al., 1999). Two major mutations, the nonsense mutation Arg208Stop and the mutation affecting splicing, IVS5-1G→C, account for approximately 60% of CLN2 chromosomes (Sleat et al., 1999; Zhong et al., 1998a); 10 cases of variant LINCL were evaluated at IBR.

Of interest, four probands with TPP 1 gene mutations demonstrated juvenile onset (4–7 years) of the disease and atypical clinical presentation. Ten NCL probands with late-infantile onset did not show mutations in the CLN2/TPP 1 gene, had normal TPP 1 levels, and showed mixed profiles at the EM level (FP, CV, RL, and GROD). These probands also were excluded from CLN2, CLN5, and CLN8. Studies are in progress to identify genetic defect(s) in this group of cases. Short summaries of the clinicopathological data for probands assigned to CLN2, but presenting with juvenile onset of the disease process, are given below.

Family 1: A family with three affected children. The onset of symptoms occurred between six and eight years of age, with behavioral abnormalities associated with dementia, coordination problems, and extrapyramidal signs. Seizures started at 9 years of age in case 1, 20 years in case 2, and 13 years in case 3. Visual impairment was not observed in case 1, who died at 24 years of age, or in case 2, who died at 30 years of age. However, visual impairment developed in case 3 at 12 years of age, leading to blindness at 20 years of age. Presently, case 3 is 28 years old and confined to a wheelchair. Buffy coat and skin-punch biopsies from cases 2 and 3 showed mainly CV profiles. Increased level of subunit c of mitochondrial ATP synthase in the urine was present in cases 2 and 3. At postmortem examination, cases 1 and 2 revealed a predominance of mixed profiles (CV and FP, GROD) in lysosomal storage material (Figure 1.10).

Studies for CLN3 mutations were negative. The activity of tripeptidyl-peptidase I (TPP1) was decreased in case 3, and molecular genetic studies showed that this proband had a common mutation leading to aberrant splicing, IVS5-1G→C, in one allele, and a missense mutation, Arg447His, in the other allele. The pedigree for this family is presented in Figure 1.11.

Proband 2: A Caucasian male had onset of symptoms at 8 years of age, with progressive blindness, seizures, cognitive and motor dysfunction, similar to family 1, case 3. He died at 43 years of age, and EM showed mixed (CV and FP) profiles in the buffy coat and in the brain. Molecular genetic studies showed the Arg208Stop nonsense mutation in one allele and a missense mutation, Arg447His, in the other allele.

V. CLN3. JUVENILE NCL

The incidence of the juvenile form of NCL varies in different countries; e.g., in Iceland it is 7.0 per 100,000 live births (Uvebrant and Hagberg, 1997), whereas

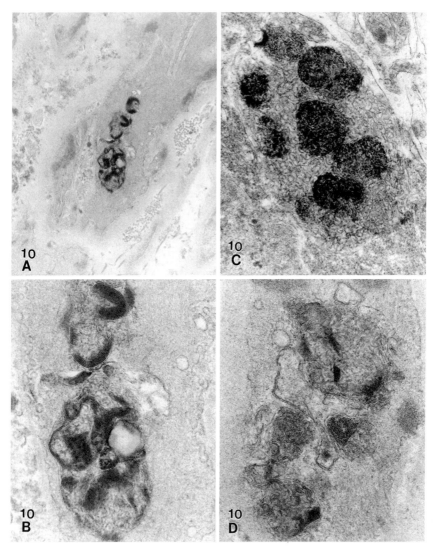

Figure 1.10. (A) Brain of CLN2 case. Mixed profiles (fingerprint and curvilinear) in the cytoplasm of a neuron, × 21,580. (B) Higher magnification of (A), × 83,000. (C) Lysosomal storage composed of mixed profiles (densely packed curvilinear and GROD), × 46,300. (D) A neuron with lysosomal storage composed of mixed profiles (rectilinear, curvilinear, and fingerprint), × 41,000.

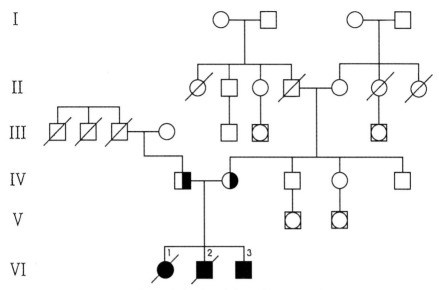

Figure 1.11. Pedigree of a family with three siblings affected with CLN2.

in West Germany it is 0.71 per 100,000 live births (Claussen et al., 1992). Forty cases evaluated at IBR were assigned to CLN3.

The course of the disease can be divided into two categories: classical/typical, and delayed classical/atypical. In all CLN3 subjects, clinical signs and symptoms appear after the age of 4 years. Rapidly progressing visual loss represents the first, and for 2–4 years, the only clinical sign of the disease. Affected children become blind usually within 2–4 years. Ophthalmological studies show pigmentary degeneration and optic nerve atrophy. Speech disturbances (frequent echolalia) and a gradual decline of cognitive functions follow after the age of 8–9 years. Epilepsy with generalized tonic-clonic seizures, myoclonic seizures, or complex partial seizures appears at the age of 5–18 years. Psychiatric signs, sleep disturbances, and extrapyramidal signs fulfill the clinical picture in the second decade of life.

EEG is abnormal, showing spike-and-slow-wave complexes. ERG and VEP are changed early. However, cranial computer tomography (CT) and MRI reveal cerebral and, to a lesser degree, also cerebellar atrophy only in the later stages of the disease.

FP are typically observed on biopsy (lymphocytes, skin, conjunctiva, rectum) and autopsy material. Vacuolated lymphocytes are a characteristic morphological feature for this form of NCL and can be found on a regular blood smear as well as in buffy coat by electron microscopy. Biochemical assays are currently unavailable. Some CLN3 subjects show increased levels of subunit c in urine samples (Wisniewski et al., 1994, 1995). Molecular genetic studies disclosed 25 different mutations and two polymorphisms in the CLN3 gene (IDBC 1995; Munroe et al., 1997; Wisniewski et al., 1998a; Zhong et al., 1998; review: Mole et al., 1999; Lauronen et al., 1999; see also Chapter 5). The major mutation is a 1.02-kb deletion, which removes exons 7 and 8. This mutation was found in 96% of JNCL patients in at least one chromosome (on around 85% of all disease chromosomes examined). The majority of homozygotes for the common deletion show a classical clinicopathological presentation with some small differences in disease progression (Järvelä et al., 1997; Lauronen et al., 1999). However, marked clinicopathological heterogeneity may be observed in compound heterozygotes.

Of the group of 40 CLN3 probands analyzed at IBR, 28 were homozygous and 12 were heterozygous for the 1.02-kb deletion. The classical clinicopathological presentation was observed in 34 probands (28/40 homozygotes and 12/40 heterozygotes). A delayed or atypical clinicopathological picture was found in six probands, all heterozygous for the 1.02-kb deletion. The age at onset for the homozygotes was 4–10 years (mean value 5.2), and for the 12 heterozygotes, 3.5–9 years (mean value 5.8). The present age of homozygotes with the 1.02-kb deletion (28/40) is 9–32 years (mean value 16.2) and for heterozygotes (12/40), 10–42 years (mean value 22.9). The age at death for homozygotes (8/28) is 18–38 years (mean value 23.5), and for heterozygotes (6/12), 24–56 years (mean value 30.6).

In two heterozygotes with the 1.02-kb deletion, we found a point mutation, the Gly295Lys (Wisniewski et al., 1998a). These cases had a protracted clinical course, with progressive visual loss that started around the age of 5 years and led to blindness after the age of 12 years. However, neurological alterations other than blindness (cognitive/motor, epilepsy) appeared in one of these cases only after the age of 40 years. The other sibling is still free of other symptoms, except for blindness at the age of 40 years. Of interest, a patient reported by Lauronen and colleagues (1999), heterozygous for the 1.02-kb deletion and the Gly295Lys mutation, also had visual loss from the age of 6 years, but afterwards developed epilepsy (since the age of 19 years) and polyneuropathy (at the age of 20 years). The cause of these phenotypic differences is unclear. It cannot be excluded that they can be attributed at least partially to environmental factors. However, in this patient, no intellectual decline or speech abnormalities could be detected (Lauronen et al., 1999). Other patients with the protracted course of the disease (Goebel et al., 1976) and with a rare pigment variant of NCL (review: Goebel et al., 1995) still have to be determined. The different diagnoses of NCL with juvenile onset are summarized in Table 1.4.

Table 1.4. NCL with Juvenile Onset >4–5 Years of Age

Gene code	Symptoms	EM	ERG/VEP	EEG	MRI/CT	Reference/mutation
CLN1	Similar to CLN3	GROD	Similar to CLN3	Similar to CLN3	Similar to CLN3	Chapter 3, *this volume*
CLN2	Similar to CLN3[a]	CV or mixed	Similar to CLN3	Similar to CLN3	Similar to CLN3	Wisniewski *et al.*, 1999, 2000
CLN3 homozygous	Blindness Dementia Seizures	FP, mixed with membrane-bound vacuoles	ERG: decreased b-wave VEP: marked reduced amplitude	Large amplitude, spikes, and slow-wave complexes	Progressive atrophy, abnormal periventricular white matter, thalamus, caudate, putamen on T2 weighted images	1.02-kb deletion Intron 6-8
Compound heterozygous	Rigidity Schuffling gait Sleep disturbances Spasticity Vegetative stage					IBDC, 1995 Chapter 5, *this volume*
Protracted	Blindness only, after 25–35 yr, other neurological abnormalities	Same as above	Same as above			Goebel *et al.*, 1993 Wisniewski *et al.*, 1998a

Juvenile onset of CLN1 age of death 15–20 yr.
[a]CLN2 onset variant longer survival, 30–43 yr; CLN3 age of death range 8–51 yr.

VI. CLN4. ADULT NCL

The adult form of NCL (CLN4, Kufs disease) can be inherited in either autosomal-recessive or dominant mode. However, because the genotype of this form is still unknown, diagnosis is based on clinicopathological data only (Berkovic et al., 1988; Goebel and Braak 1989; Constantinidis et al., 1992; Martin and Ceuterick 1997).

Initial signs and symptoms appear usually around the age of 30 years; however, they may appear as early as the age of 11 years. There are two major clinical phenotypes. Type A is characterized by progressive myoclonic epilepsy with dementia, ataxia, and late-occurring pyramidal extrapyramidal signs. Affected individuals usually have no visual problems. Seizures often are uncontrollable. Type B is characterized by behavior abnormalities and dementia, which may be associated with motor dysfunction and disturbances from the cerebellum (ataxia) and basal ganglia (extrapyramidal signs). Dysarthria and suprabulbar signs are also features of this disorder (Berkovic et al., 1988). In the presenile form, onset after 50 years of age, dementia, cognitive decline, motor dysfunction, seizures, and suprabulbar signs (brain stem) are present (Constantinidis et al., 1992). Some probands may have A and B phenotype (Martin et al., 1987). Literature on the adult form of NCL is reviewed by Berkovic et al. (1988) and by Constantinidis et al. (1992).

Electrophysiological and neuroradiological studies may be helpful for obtaining the final diagnosis, but their findings are not specific. Ophthalmological studies are normal, in contrast to studies in childhood forms of NCLs. Thus, demonstration of lysosomal inclusion bodies that are usually mixed (FP, GROD, RL, or CV) profiles in biopsy material (skin, conjunctive, muscle, rectal), and if necessary, in brain biopsy, is mandatory for definitive diagnosis (Martin et al., 1987; Martin and Ceuterick, 1997).

VII. CLN5. FINNISH LATE-INFANTILE VARIANT

The Finnish late-infantile variant is highly enriched in that country, where it accounts for 80% of late-infantile NCL cases reported (Santavouri et al., 1982; Uvebrant and Hagberg 1997). The CLN5 gene was assigned to chromosome 13q22 (Savukoski et al., 1994) and subsequently isolated and characterized. Details are described by Ranta et al. (see Chapter 6).

The major features differentiating CLN5 subjects from subjects with classical CLN2 are later onset of the disease (usually 4.5–7 years) and the presence of FP and RL profiles apart from CV profiles at the EM level. Lymphocytes are not vacuolated. Accumulation of subunit c of mitochondrial ATP synthase and small amounts of saposin A and D are present (Tyynelä et al., 1997). Thus, these cases resemble classical late-infantile NCL in their clinical features, but the onset of symptoms is at a later age, and the EM findings are different than in classical CLN2. Presently, three mutations of CLN5 and one polymorphism are known. All probands are of Finnish origin, except one Dutch and Swedish (Savukoski *et al.*, 1998). The gene product of CLN5 is a transmembrane protein of unknown function that is similar to CLN3 protein (Savukoski *et al.*, 1998). Biochemical diagnostic assays are not available.

VIII. CLN6. VARIANT LATE-INFANTILE GYPSY/INDIAN

The incidence of the late-infantile/early-juvenile variant has not yet been determined. The CLN6 variant appears to be especially common in some European countries and Canada (Lake and Cavanaugh, 1978; Andermann *et al.*, 1988; Elleder *et al.*, 1997, 1999). CLN6 has been mapped to chromosome 15q21-23 (Sharp *et al.*, 1997), but has not yet been isolated or characterized. Of the families tested for linkage to CLN6, because of similarities in clinical features and ultrastructural appearances (FP, rarely CV or RL, profiles without vacuoles), only approximately 75% showed evidence of linkage, confirming that locus genetic heterogeneity exists (Williams *et al.*, 1999a).

A series of 27 Czech cases from 23 families has been described, of which 25 cases were diagnosed while living (Elleder *et al.*, 1997, 1999). The age at onset ranged from 18 months to 8 years. Genetic analysis has allowed another 31 families to be assigned to CLN6 given current linkage data (Williams *et al.*, 1999a). In less than 50% of variant LINCL (CLN6) cases, clinical presentation is similar to that in classical LINCL (CLN2), but mostly RL and FP profiles are found ultrastructurally, although CV profiles also may occur (Figures 1.12 and 1.13).

Visual loss and seizures may represent the initial clinical signs and symptoms. However, in some children with later onset of the disease process (after the age of 4 years), the disease may start with epilepsy, ataxia, and myoclonus (Williams *et al.*, 1999a). EEG, ERG, VEP, and MRI are helpful in establishing diagnosis. Electron microscopy studies of skin, rectum, or lymphocytes are necessary. However, definitive diagnosis should be based on molecular genetic and biochemical studies, which are not yet available.

IX. CLN7. TURKISH VARIANT LATE-INFANTILE NCL

CLN7 is a Turkish variant of late-infantile NCL. Between 1993 and 1997, seven of eight cases in Ankara, Turkey, have been diagnosed, with onset at 1–6 years of age; most families of these cases were consanguineous. Presently, 14 cases of Turkish origin have been identified. They have a similar phenotype to CLN2 but

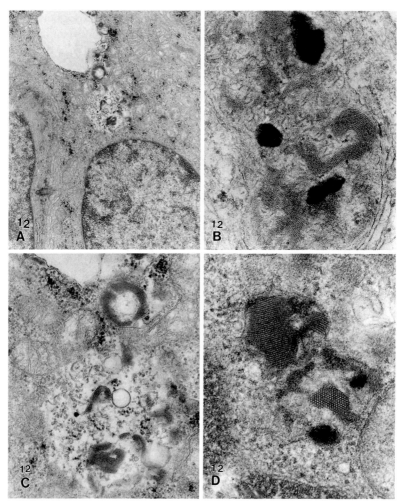

Figure 1.12. (A) CLN6 variant LINCL. In the secretory exocrine sweat glands, epithelial cell from a skin punch biopsy; lysosomal storage material composed of mixed profiles (fingerprint, trilamellar structures), × 18,400. (B) Higher magnification of (A), × 55,000. (C) Fibroblasts with rectilinear complex (fingerprints, trilamellar, and angulated) fibers, × 175,000. (D) Fibroblast with island of fingerprints and honeycomb-like structures in lysosomal storage material, without membrane-bound vacuoles, × 190,000.

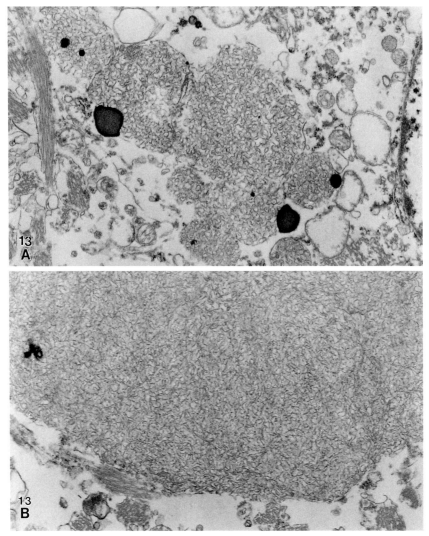

Figure 1.13. Brain of CLN6 case. (A) Membranous profiles (rectilinear complexes) with slightly wavy course at various lengths and bending patterns; some superficially resembling curvilinear profiles typical of earlier juvenile CLN6 (Czech late-infantile variant of NCL), × 52,000. (B) Lower magnification of (A), × 44,500.

Table 1.5. Late-Infantile Forms of NCL

Form	Gene code	Onset (yr)	First symptom	EM	Enzyme level	References
LINCL	CLN1	2–4	Motor dysfunction	GROD	PPT1 deficiency	Chapter 3, *this volume*
Classical LINCL	CLN2	2–8	Seizures	CV, mixed	TPP1 deficiency	Sleat *et al.*, 1999; Goebel *et al.*, 1999
Finnish LINCL	CLN5	4–7	Motor dysfunction	FP or mixed[a]	TPP1 normal	Chapter 6, *this volume*
LINCL "early juvenile"	CLN6	1.5–8	Seizure	Mixed[a]	TPP1 normal	Lake and Cavanagh, 1978
Turkish LINCL	CLN7	1.0–6	Seizures	CV, GROD-like structures, mixed	TPP1 normal	Williams *et al.*, 1999a, 1996b

[a]Without membrane-bound vacuoles. Age of death of CLN1, CLN2, CLN5 7:4·25–43:25 yr.

have earlier or later age at onset, and at the EM level, showed FP and/or mixed profiles (FP, CV, RL). To date, CLN7 has been excluded from the known NCL gene loci. These exclusion data eliminate the possibility that CLN7 is an allelic variant of one of the other NCLs (Williams et al., 1999b).

The different diagnoses of all NCLs with late-infantile onset are summarized in Table 1.5.

X. CLN8. NORTHERN EPILEPSY

Detailed clinicopathologic and molecular genetic data about the northern epilepsy variant of NCL are presented by Ranta et al. (see Chapter 6). Clinical onset occurs at 5–10 years. Diagnosis is based on a characteristic clinical picture (epilepsy with tonic-clonic or complex partial seizures, mental deterioration, motor dysfunction) and EM studies showing storage material that resembles CV profiles and GROD. Definitive diagnosis is based on molecular genetic studies. The CLN8 gene was isolated and characterized recently (Ranta et al., 1999). All 22 patients examined were homozygous to a missense mutation, Arg24Gly. Carrier frequency for the CLN8 gene is 1:135.

XI. SUMMARY

Previously, before the era of genetic studies, diagnosis of NCL was based on clinicopathological findings alone, and four forms of NCL were known. No explanation was given for atypical NCL cases (Wisniewski et al., 1988, 1992; Dyken and Wisniewski, 1995) that showed different clinicopathological presentation from the four typical forms. Presently, NCLs are classified into a total of eight forms on the basis of clinicopathological and genetic findings. The data presented above show that although only biochemical and molecular genetic studies allow for definitive diagnosis, ultrastructural studies of the biopsy material are still very useful. Furthermore, the identification of numerous mutations in each of the major NCL groups in children presented difficulties in assigning new cases to a particular NCL form. Thus, to facilitate clinical diagnosis, biochemical assays based on measurements of activity of PPT (Verkruyse et al., 1997; Cho and Dawson 1998) and PPT1 (Junaid et al., 1999; Sohar et al., 1999) have been developed recently. Further studies aimed at characterizing the structural and functional properties of the NCL genes' products are under way. Studies also are in progress to elucidate the influence of mutations on the properties of NCL-associated gene products to better understand the pathogenesis and phenotypic heterogeneity of these disorders. Further research should better define the pathogenesis of the NCLs to determine the genetic defect in the variant late-infantile (CLN6–7) and the adult (CLN4) forms, to determine the enzyme defect in all forms, and to develop treatment by enzyme or gene replacement and possible pharmaceutical agents (see Chapter 12). [Continue role of genetic counseling, (see chapter 8) by S. Sklower Brooks.]

Acknowledgments

We wish to thank the families and their referring physicians who have participated in these studies; the Batten Disease Support and Research Association (B.D.S.R.A.), Lance Johnston, Executive Director, and Larry Killen, President; and Edith Dockter, coordinator of the Batten Disease Registry at IBR. Supported in part by a grant from the National Institutes of Health, National Institute of Neurological Disorders and Stroke, NS/HD38988, and the New York State Office of Mental Retardation and Developmental Disabilities (OMRDD).

References

Andermann, E., Jacob, J. C., Andermann, F., Carpenter, S., Wolfe, L., and Berkovic, S. F. (1988). The Newfoundland aggregate of neuronal ceroid-lipofuscinosis. *Am. J. Med. Genet. Suppl.* **5,** 111–116.

Batten, F. E. (1903). Cerebral degeneration with symmetrical changes in the maculae in two members of the family. *Trans. Ophthalmol. Soc. UK* **23,** 386–390.

Berkovic, S. F., Carpenter, S., Andermann, F., Andermann, E., and Wolfe, L. S. (1988). Kufs' disease: A critical reappraisal. *Brain* **11,** 27–62.

Bielschowsky, M. (1923). Über späte-infantile familiäre amaurotische Idiotie mit Kleinhimsymptomen. *Dtsch. Z. Nervenheilk.* **50,** 7–29.

Camp, L. A., and Hofmann, S. L. (1993). Purification and properties of a palmitoyl-protein thioesterase that cleaves palmitat from H-Ras. *J. Biol. Chem.* **268,** 22566–22574.

Camp, L. A., Verkruyse, L. A., Afendit, C. A., Slaughter, C. A., and Hofmann, S. L. (1994). Molecular cloning and expression of palmitoyl-protein thioesterase. *J. Biol. Chem.* **269,** 23212–23219.

Cardona, F., and Rosati, E. (1995). Neuronal ceroid-lipofuscinoses in Italy: An epidemiological study. *Am. J. Med. Genet.* **57,** 142–143.

Carpenter, S., Karpati, G., Wolfe, L. S., and Andermann, F. (1973). A type of juvenile cerebromacular degeneration characterized by granular osmiophilic deposits. *J. Neurol. Sci.* **18,** 67–87.

Cho, S., and Dawson, G. (1998). Enzymatic and molecular biological analysis of palmitoyl protein thioesterase deficiency in infantile neuronal ceroid lipofuscinosis. *J. Neurochem.* **71,** 323–329.

Claussen, M., Heim, P., Knispel, J., Goebel, H. H., and Kohlschütter, A. (1992). Incidence of neuronal ceroid-lipofuscinoses in West Germany: Variation of a method for studying autosomal recessive disorders. *Am. J. Med. Genet.* **42,** 536–538.

Constantinidis, J., Wisniewski, K. E., and Wisniewski, T. M. (1992). The adult and a new late forms of neuronal ceroid lipofuscinosis, *Acta Neuropathol. (Berl.)* **83,** 461–468.

Chow, C. W., Borg, J., Billson, V. R., and Lake, B. D. (1993). Fetal tissue involvement in the late infantile type of neuronal ceroid lipofuscinosis. *Prenat. Diagn.* **13,** 833–841.

Crow, Y. J., Tolmie, J. L., Howatson, A. G., Patrick, W. J., and Stephenson, J. B. (1997). Batten disease in the West of Scotland 1974–1995 including five cases of the juvenile form with granular osmiophilic deposits. *Neuropediatrics* **28,** 140–144.

Das, A. K., Becerra, C. H. R., Yi, W., Lu, J.-Y., Siakotos, A. N., Wisniewski, K. E., and Hofmann, S. L. (1998). Molecular genetics of palmitoyl-protein thioesterase deficiency in the US. *J. Clin. Invest.* **101,** 361–370.

Dyken, P., and Wisniewski, K. (1995). Classification of the neuronal ceroid-lipofuscinoses: Expansion of the atypical forms. *Am. J. Med. Genet.* **57,** 150–154.

Elleder, M., Franc, J., Kraus, J., Nevsimalova, S., Sixtova, K., and Zeman, J (1997). Neuronal ceroid lipofuscinosis in the Czech Republic: Analysis of 57 cases. Report of the 'Prague NCL group'. *Eur. J. Pediatr. Neurol.* **4,** 109–114.

Elleder, M., Kraus, J., Nevsimalová, S., Sixtová, K., and Zeman, J. (1999). NCL in different European countries: Czech Republic. *In:* "The Neuronal Ceroid Lipofuscinoses (Batten Disease)" (H. H. Goebel *et al.*, eds.), p. 129. IOS Press, Amsterdam, The Netherlands.

Goebel, H. H., and Braak, H. (1989). Adult neuronal ceroid lipofuscinosis. *Clin. Neuropathol.* **8,** 109–119.

Goebel, H. H., Pilz, H., and Gullotta, F. (1976). The protracted form of juvenile neuronal ceroid lipofuscinosis. *Acta Neuropathol. (Berl.)* **36,** 393–396.

Goebel, H. H., Gullotta, F., Bajanowski, T., Hansen, F. J., and Braak, H. (1995). Pigment variant of neuronal ceroid-lipofuscinosis. *Am. J. Med. Genet.* **57,** 155–159.

Goebel, H. H., Mole, S. E., and Lake, B. D. (1999). Introduction. *In:* "The Neuronal Ceroid Lipofuscinoses (Batten Disease)" (H. H. Goebel *et al.*, eds.), pp. 1–4. IOS Press, Amsterdam, The Netherlands.

Hallervorden, J. (1938). Spätform der amaurotischen Idiotie unter Bilde der Paralysis agitans. *Mschr. Psychiat. Neurol.* **99,** 74–80.

Haltia, M., Rapola, J., and Santavouri, P. (1973). Infantile type of so-called neuronal ceroid lipofuscinosis. Histological and electron microscopic studies *Acta Neuropathol. (Berl.)* **26,** 157–170.

Hartikainen, J. M., Ju, W., Wisniewski, K. E., Moroziewicz, D. N., Kaczmarski, A. L., McLendon, L., Zhong, D., Suarez, C. T., Brown, W. T., and Zhong, N. (1999). Late infantile neuronal ceroid lipofuscinosis is due to splicing mutations in the CLN2 gene. *Mol. Genet. Metab.* **67,** 162–168.

Hofman, I. L., and Taschner, E. M. (1995). Late onset juvenile neuronal ceroid-lipofuscinosis with granular osmiophilic deposits (GROD). *Am. J. Med. Genet.* **57,** 165–167.

Hofmann, S. L., Das, A. K., Yi, W., Lu, J-Y., and Wisniewski, K. E. (1999). Genotype-phenotype correlations in neuronal ceroid lipofuscinosis due to palmitoyl-protein thioesterase deficiency. *Mol. Genet. Metab.* **66,** 234–239.

International Batten Disease Consortium (IBDC) (1995). Isolation of a novel gene underlying Batten disease, CLN3. *Cell* **82,** 949–957.

Janes, R. W., Munroe, P. B., Mitchison, H. M., Gardiner, R. M., Mole, S. E., and Wallace, B. A. (1996). A model for Batten disease protein CLN3: Functional implications from homology and mutations. *FEBS Lett.* **399,** 75–77.

Janský, J. (1908). Dosud nepopsaný případ familiární aurotické-idiotie kompliklované s hypoplasií mozečkovou. *Sborn. Lék.* **13,** 165–196.

Järvelä, I., Autti, T., Lamminranta, S., Aberg, L., Raininko, R., and Santavuori, P. (1997). Clinical and magnetic resonance imaging findings in Batten disease: Analysis of the major mutation (1.02-kb deletion). *Ann. Neurol.* **42,** 799–802.

Junaid, M. A., Sklower Brooks, W., Wisniewski, K. E., and Pullarkat, R. K. (1999). A novel assay for lysosomal pepstatin-insensitive protease and its application for the diagnosis of late-infantile neuronal ceroid lipofuscinosis. *Clin. Chim. Acta* **281,** 169–176.

Kufs, H. (1925). Über eine Spätform der amaurotischen Idiotie und ihre heredofamiliaren Grundlagen. *Z. Ges. Neurol. Psychiat.* **95,** 169–188.

Lake, B. D., and Cavanaugh, N. P. (1978). Early-juvenile Batten's disease—A recognizable sub-group distinct from other forms of Batten's disease. Analysis of 5 patients. *J. Neurol. Sci.* **36,** 265–271.

Lake, B. D., Brett, E. M., and Boyd, S. G. (1996). A form of juvenile Batten disease with granular osmiophilic deposits. *Neuropediatrics* **27,** 265–269.

Lake, B. D., Steward, C. G., Oakhill, A., Wilson, J., and Perham, T. G. M. (1997). Bone marrow transplantation in late infantile Batten disease and juvenile Batten disease. *Neuropediatrics* **28,** 80–81.

Lauronen, L., Munroe, P. B., Jarvela, I., Autti, T., Mitchison, H. M., O'Rawe, A. M., Gardiner, R. M., Mole, S. E., Puranen, J., Hakkinen, A. M., Kirveskari, E., and Santavuori, P. (1999). Delayed classic and protracted phenotypes of compound heterozygous juvenile neuronal ceroid lipofuscinosis. *Neurology* **52,** 360–365.

Manca, V., Kanitakis, J., Zambruno, G., Thivolet, J., and Gonnaud, P. M. (1990). Ultrastructural study of the skin in a case of juvenile ceroid-lipofuscinosis. *Am. J. Dermatopathol.* **12,** 412–416.

Martin, J. J., and Ceuterick, C. (1997). Adult neuronal ceroid-lipofuscinosis: Personal observations. *Acta Neurol. Belg.* **97,** 85–92.

Martin, J. J., Libert, J., and Ceuterick, C. (1987). Ultrastructure of brain and retina in Kufs'disease (adult type-ceroid-lipofuscinosis). *Clin. Neuropathol.* **6,** 231–235.

Mitchison, H. M., Hofmann, S. L., Becerra, C. H., Munroe, P. B., Lake, B. D., Crow, Y. J., Stephenson, J. B., Williams, R. E., Hofman, I. L., Taschner, P. E. M., Martin, J. J., Philippart, M., Andermann, E., Andermann, F., Mole, S. E., Gardiner, R. M., and O'Rawe, A. M. (1998). Mutations in the palmitoyl-protein thioesterase gene (PPT: CLN1) causing juvenile neuronal ceroid lipofuscinosis with granular osmiophilic deposits. *Hum. Mol. Genet.* **7,** 291–297.

Mole, S. E., Mitchison, H. M., and Munroe, P. B. (1999). Molecular basis of the neuronal ceroid lipofuscinoses: Mutations in CLN1, CLN2, CLN3, and CLN5. *Hum. Mutat.* **14,** 199–215.

Munroe, P. B., Mitchison, H. M., O'Rawe, A. M., Anderson, J. W., Boustany, R-M., Lerner, T. J., and Taschner, P. E. M. (1997). Spectrum of mutations in the Batten disease gene, CLN3. *Am. J. Hum. Genet.* **61,** 310–316.

Philippart, M., Chugani, H. T., and Bronwyn Bateman, J. (1995). New Spielmeyer-Vogt variant with granular inclusions and early brain atrophy. *Am. J. Med. Genet.* **57,** 160–164.

Ranta, S., Zhang, Y., Ross, B., Lonka, L., Takkunen, E., Messer, A., Sharp, J., Wheeler, R., Kusumi, K., Mole, S., Liu, W., Soares, M. B., de Fatima Bonaldo, M., Hirvasniemi, A., de la Chapelle, A., Gilliam, T. C., and Lehesjoki, A-E. (1999). The neuronal ceroid lipofusinoses in human EPMR and mnd mutant mice are associated with mutations in CLN8. *Nat. Genet.* **23,** 233–236.

Rapola, J., Salonen, R., Ammala, P., and Santavuori, P. (1990). Prenatal diagnosis of the infantile type of neuronal ceroid lipofuscinosis by electron microscopic investigation of human chorionic villi. *Prenat. Diagn.* **10,** 553–559.

Rawlings, N. D., and Barrett, A. J. (1999). Tripeptidyl-peptidase I is apparently the CLN2 protein absent in classical late-infantile neuronal ceroid lipofuscinosis. *Biochem. Biophys. Acta* **1429,** 496–500.

Santavuori, P. (1988). Neuronal ceroid-lipofuscinoses in childhood. *Brain Dev.* **10,** 80–83.

Santavuori, P., Haltia, M., Rapola, J., and Raitta, C. (1973). Infantile type of so-called neuronal ceroid-lipofuscinosis Part 1. A clinical study of 15 patients. *J. Neurol. Sci.* **18,** 257–267.

Santavuori, P., Rapola, J., Sainio, K., and Raitta, C. (1982). A variant of Jansky-Bielchowsky disease. *Neuropediatrics* **13,** 135–141.

Santavuori, P., Heiskala, H., Autti, T., Johansson, E., and Westermarck, T. (1989). Comparison of the clinical courses in patients with juvenile neuronal ceroid lipofuscinosis receiving antioxidant treatment and those without antioxidant treatment. In: "Lipofuscin and Ceroid Pigments" (E. A. Porta, ed.), pp. 273–282. Plenum Press, New York.

Santavuori, P., Gottlob, I., Haltia, M., Rapola, J., Lake, B. D., Tyynelä, J., and Peltonen, L. (1999). CLN1 Infantile and other types of NCL with GROD. In: "The Neuronal Ceroid Lipofuscinoses (Batten Disease)" (H. H. Goebel et al., eds.), pp. 16–36. IOS Press, Amsterdam, The Netherlands.

Santorelli, F. M., Bertini, E., Petruzzella, V., Di Capua, M., Calvieri, S., Gasparini, P., and Zeviani, M. (1998). A novel insertion mutation (A169I) in the CLN1 gene is associated with infantile neuronal ceroid lipofuscinosis in an Italian patient. *Biochem. Biophys. Res. Commun.* **245,** 519–522.

Savukoski, M., Kestilä, M., Williams, R., Järvelä, I., Sharp, J., Harris, J., Santavuori, P., Gardiner, M., and Peltonen, L. (1994). Defined chromosomal assignment of CLN5 demonstrates that at least

four genetic loci are involved in the pathogenesis of human ceroid lipofuscinosis. *Am. J. Hum. Genet.* **55,** 695–701.
Savukoski, M., Klockars, T., Holmberg, V., Santavuori, P., Lander, E. S., and Peltonen, L. (1998). CLN5, a novel gene encoding a putative transmembrane protein mutated in Finnish variant late infantile neuronal ceroid lipofuscinosis. *Nat. Genet.* **19,** 286–288.
Sharp, J. D., Wheeler, R. B., Lake, B. D., Savukowski, M., Järvelä, I. E., Peltonen, L., Gardiner, R. M., and Williams, R. E. (1997). Loci for classical and a variant late infantile neuronal ceroid lipofuscinosis map to chromosomes 11p15 and 15q21-23. *Hum. Mol. Genet.* **6,** 591–595.
Sleat, D. E., Donnelly, R. J., Lackland, H., Liu, C. G., Sohar, I., Pullarkat, R. K., and Lobel, P. (1997). Association of mutations in a lysosomal protein with classical late-infantile neuronal ceroid lipofuscinosis. *Science.* **277,** 1802–1805.
Sleat, D. E., Gin, R. M., Sohar, I., Wisniewski, K., Sklower-Brooks, S., Pullarkat, R. K., Palmer, D. N., Lerner, T. J., Boustany, R-M., Uldall, P., Siakotos, A. N., Donnelly, R. J., and Lobel, P. (1999). Mutational analysis of the defective protease in classical late-infantile neuronal ceroid lipofuscinosis, a neurodegenerative lysosomal storage disorder. *Am. J. Hum. Genet.* **64,** 1511–1523.
Sohar, I., Sleat, D. E., Jadot, M., and Lobel, P. (1999). Biochemical characterization of a lysosomal protease deficient in classical late infantile neuronal ceroid lipofuscinosis (LINCL) and development of an enzyme-based assay for diagnosis and exclusion of LINCL in human specimens and animal models. *J. Neurochem.* **73,** 700–711.
Spielmeyer, W. (1923). Familiare amaurotische idotie. *Zentralbl. Gesamte. Ophthal.* **10,** 161–208.
Stengel, C. (1826). Account of a singular illness among four siblings in the vicinity of Røraas. *In:* "Ceroid-Lipofuscinosis (Batten's Disease)" (D. Armstrong *et al.*, eds.), pp. 17–19 Elsevier Biomedical Press, Amsterdam, The Netherlands, 1982. Translated from: Beretning om et maerkeligt Sygdomstilfoelde hos fire Sødskende I Naeheden af Røraas, *Eyr et medicinsk Tidskrift.* **1,** 347–352.
Stephenson, J. B. P., Greene, N. D. E., Leung, K. Y., Munroe, P. B., Mole, S. E., Gardiner, R. M., Taschner, P. E. M., O'Regan, M., Naismith, K., Crow, Y. J., and Mitchison, H. M. (1999). The molecular basis of GROD-storing neuronal ceroid lipofuscinoses in Scotland. *Mol. Genet. Metab.* **66,** 245–247.
Taratuto, A. L., Saccoliti, M., Sevlever, G., Ruggieri, V., Arroyo, H., Herrero, M., Massaro, M., and Fejerman, N. (1995). Childhood neuronal ceroid-lipofuscinoses in Argentina. *Am. J. Med. Genet.* **57,** 144–149.
Tyynelä, J., Suopanki, J., Santavuori, P., Baumann, M., and Haltia, M. (1997). Variant late infantile neuronal ceroid-lipofuscinosis: Pathology and biochemistry. *J. Neuropath. Exp. Neurol.* **56,** 369–375.
Uvebrant, P., and Hagberg, B. (1997). Neuronal ceroid lipofuscinosis in Scandinavia. Epidemiological and clinical pictures. *Neuropediatrics* **28,** 6–8.
Van Diggelen, O. P, Keulemans, J. L. M., Winchester, B., Hofman, I. L., Vanhanen, S. L., Santavuori, P., and Voznyi, Y. V. (1999). A rapid fluorogenic palmitoyl-protein thioesterase assay: Pre-and postnatal diagnosis of INCL. *Mol. Genet. Metab.* **66,** 240–244.
Verkruyse, L. A., Natowicz, M. R., and Hofmann, S. L. (1997). Palmitoyl-protein thioesterase deficiency in fibroblasts of individuals with infantile neuronal ceroid lipofuscinosis and I-cell disease. *Biochem. Biophys. Acta* **1361,** 1–5.
Vesa, J., Hellsten, E., Verkruyse, L. A., Camp, L. A., Rapola, J., Santavuori, P., Hofmann, S. L., and Peltonen, L. (1995). Mutations in the palmitoyl-protein thioesterase gene causing infantile neuronal ceroid lipofuscinosis. *Nature.* **376,** 584–587.
Vines, D., and Warburton, N. J. (1999). Classical late infantile neuronal ceroid lipofuscinosis fibroblasts are deficient in lysosomal tripeptidyl peptidase I. *FEBS Lett.* **443,** 131–135.

Waliany, S. A. K., Das, A. K., Gaben, A., Wisniewski, K. E., and Hofmann, S. L. (1999). Identification of three novel mutations of the palmitoyl-protein thioesterase (PPT) gene in children with neuronal ceroid-lipofuscinosis. *Hum. Mutat.*, Mutation in Brief #290.

Williams, R. E., Lake, B. D., Elleder, M., and Sharp, J. D. (1999a). CLN6 Variant Late Infantile/Early Juvenile NCL. *In:* "The Neuronal Ceroid Lipofuscinoses (Batten Disease)" (H. H. Goebel *et al.*, eds.), pp. 102–116 IOS Press, Amsterdam, The Netherlands.

Williams, R. E., Topçu, M., Lake, B. D., Mitchell, W., and Mole, S. E. (1999b). CLN7 Turkish Variant Late Infantile NCL. *In:* "The Neuronal Ceroid Lipofuscinoses (Batten Disease)" (H. H. Goebel *et al.*, eds.), pp. 114–116 IOS Press, Amsterdam, The Netherlands.

Wisniewski, K. E., Rapin, I., and Heaney-Kieras, J. (1988). Clinico-pathological variability in the childhood neuronal ceroid lipofuscinoses and new observations on glycoprotein abnormalities. *Am. J. Med. Genet. Suppl.* **5**, 27–46.

Wisniewski, K. E., Kida, E., Patxot, O. F., and Connell, F. (1992). Variability in the clinical and pathological findings in the neuronal ceroid lipofuscinoses: Review of data and observations. *Am. J. Med. Genet.* **42**, 525–532.

Wisniewski, K. E., Golabek, A. A., and Kida, E. (1994). Increased urine levels of subunit c of mitochondrial ATP synthase in neuronal ceroid lipofuscinosis patients. *J. Inher. Metab. Dis.* **17**, 205–210.

Wisniewski, K. E., Kaczmarski, W., Golabek, A. A., and Kida, E. (1995). Rapid detection of subunit c of mitochondrial ATP synthase in urine as a diagnostic screening method for neuronal ceroid-lipofuscinoses. *Am. J. Med. Genet.* **57**, 246–249.

Wisniewski, K. E., Zhong, N., Kaczmarski, W., Kaczmarski, A., Kida, E., Brown, W. T., Schwarz, K. O., Lazzarini, A. M., Rubin, A. J., Stenroos, E. S., Johnson, W. G., and Wisniewski, T. M. (1998a). Compound heterozygous genotype is associated with protracted juvenile neuronal ceroid lipofuscinosis. *Ann. Neurol.* **43**, 106–110.

Wisniewski, K. E., Zhong, N., Kaczmarski, W., Kaczmarski, A., Sklower-Brooks, S., and Brown, W. T. (1998b). Studies of atypical JNCL suggest overlapping with other NCL forms. *Ped. Neurol.* **18**, 36–40.

Wisniewski, K. E., Kida, E., Connell, F., Kaczmarski, W., Kaczmarski, A., Michalewski, M. P., and Zhong, N. (1999). Reevaluation of neuronal ceroid lipofuscinosis: Atypical juvenile onset may be the result of CLN2 mutations. *Mol. Genet. Metab.* **66**, 248–252.

Wisniewski, K. E., Kida, E., Connell, Zhong, N. (2000). Neuronal ceroid lipofuscinoses: research update. *Neurol. Sci.* **Suppl (3) 21**, S49–S56.

Zeman, W. (1976). The neuronal ceroid lipofuscinoses *In:* "Progress in Neuropathology" (H. M. Zimmerman, ed.), Vol. III, pp. 203–223. Grune & Stratton, New York.

Zeman, W., and Dyken, P. (1969). Neuronal ceroid-lipofuscinosis (Batten's disease). Relationship to amaurotic family idiocy? *Pediatrits* **44**, 570–583.

Zhong, N., Wisniewski, K. E., Hartikainen, J., Ju, W., Moroziewicz, D. N., McLendon, L., Sklower Brooks, S., and Brown, W. T. (1998a). Two common mutations in the CLN2 gene underlie late infantile neuronal ceroid lipofuscinosis. *Clin. Genet.* **54**, 234–238.

Zhong, N., Wisniewski, K. E., Kaczmarski, A. L., Ju, W., Xu, W., Xu, W. W., McLendon, L., Liu, B., Kaczmarski, W., Sklower Brooks, S., and Brown, W. T. (1998b). Molecular screening of Batten disease: Identification of a missense mutation (E295K) in the CLN3 gene. *Hum. Genet.* **102**, 57–62.

2

Cellular Pathology and Pathogenic Aspects of Neuronal Ceroid Lipofuscinoses

Elizabeth Kida, Adam A. Golabek,* and Krystyna E. Wisniewski
Department of Pathological Neurobiology
New York State Institute for Basic Research in Developmental Disabilities
Staten Island, New York 10314

    I. Introduction
   II. CLN1. Infantile Form of NCL: Deficiency of Palmitoyl-Protein Thioesterase 1
  III. CLN2. The Classic Late-Infantile NCL: Deficiency of Tripeptidyl-Peptidase I
  IV. CLN3. Juvenile Form of NCL: Genetic Defect of Lysosomal Membrane Protein
   V. Other NCL Forms with Known Genetic Defect
  VI. NCL Forms without Identified Genetic Defects
      References

ABSTRACT

Lysosomal accumulation of autofluorescent, ceroid lipopigment material in various tissues and organs is a common feature of the neuronal ceroid lipofuscinoses (NCLs). However, recent clinicopathologic and genetic studies have evidenced that NCLs encompass a group of highly heterogeneous disorders. In five of the eight NCL variants distinguished at present, genes associated with the disease process have been isolated and characterized (CLN1, CLN2, CLN3, CLN5, CLN8). Only products of two of these genes, CLN1 and CLN2, have structural and functional properties of lysosomal enzymes. Nevertheless, according to the nature of

*Address for correspondence: E-mail: golabek@popmail.med.nyu.edu the material accumulated in the lysosomes, NCLs in humans as well as natural animal models of these disorders can be divided into two major groups: those characterized by the prominent storage of saposins A and D, and those showing the predominance of subunit c of mitochondrial ATP synthase accumulation. Thus, taking into account the chemical character of the major component of the storage material, NCLs can be classified currently as proteinoses. Of importance, although lysosomal storage material accumulates in NCL subjects in various organs, only brain tissue shows severe dysfunction and cell death, another common feature of the NCL disease process. However, the relation between the genetic defects associated with the NCL forms, the accumulation of storage material, and tissue damage is still unknown. This chapter introduces the reader to the complex pathogenesis of NCLs and summarizes our current knowledge of the potential consequences of the genetic defects of NCL-associated proteins on the biology of the cell.

I. INTRODUCTION

The lysosomal storage diseases encompass at present more than 40 different genetically determined disorders, with a combined birth prevalence in the range of 14 per 100,000 live births (Poorthuis et al., 1999). Most of these disorders are caused by abnormal hydrolysis of an adequate substrate, either by mutated lysosomal hydrolase or by mutated activator or protector protein. Deficient activity of lysosomal hydrolase leads to the accumulation of undigested product in the lysosomal lumen and dysfunction of vulnerable tissue. However, in mucolipidosis II and mucolipidosis III, lysosomal storage is caused by altered targeting of lysosomal hydrolases as a result of deficient activity of GlcNAc-phosphotransferase (Reitman and Kornfeld, 1981). In nephropathic cystinosis or sialic acid storage disorders, lysosomal storage is caused by a genetic defect of lysosomal transporter proteins, cystinosin (Town et al., 1998) and sialin (Verheijen et al., 1999), respectively, both with the predicted multitransmembrane topology. Thus, pathogenic events leading to lysosomal storage may vary and include dysfunctions of both enzymes and transporter proteins.

Zeman and Dyken (1969) introduced the term neuronal ceroid lipofuscinoses (NCLs) to separate this group of disorders from the gangliosidoses, to which they were traditionally assigned. At present, NCLs encompass a group of diseases that are highly heterogeneous from both the clinicopathological and genetic points of view. A common feature of all forms of neuronal ceroid lipofuscinoses is accumulation of autofluorescent, ceroid lipopigment material in lysosomal compartments. However, data collected to the present suggest that the pathomechanisms responsible for lysosomal storage within the group of NCLs may vary. Five of eight NCL-associated genes have been cloned so far: CLN1, CLN2, CLN3, CLN5, and CLN8 (Vesa et al., 1995; Sleat et al., 1997; The International Batten

Disease Consortium, 1995; Savukoski et al., 1998; Ranta et al., 1999). Of interest, only products of two of these genes, CLN1 and CLN2, have structural and functional properties of lysosomal enzymes. The predicted amino acid composition of three other NCL-associated proteins, CLN3p, CLN5p, and CLN8p, suggests that these proteins have multiple transmembrane domains, which implies that they might not act as soluble lysosomal enzymes.

In spite of the well-documented clinicopathologic and genetic heterogeneity of NCLs, according to the nature of the material acccumulated in lysosomes, NCLs in humans as well as animal forms of these disorders can be divided into two major groups: those characterized by the prominent storage of saposins (SAPs) A and D, and those showing the predominance of subunit c of mitochondrial ATP synthase accumulation (Palmer et al., 1997a). In addition to proteins, storage material in NCLs contains other components such as lipids, metals, dolichyl pyrophosphoryl oligosaccharides, and lipid thioesters (Wolfe et al., 1977; Goebel et al., 1979; Ng Ying Kin et al., 1983; Banerjee et al., 1992; Palmer et al., 1992; Lu et al., 1996; Dawson et al., 1997).

The relation between genetic defects associated with the major NCL forms, the accumulation of storage material, and tissue dysfunction and/or damage is still unknown. Furthermore, subjects with all NCL forms manifest lysosomal storage in many tissues and organs, but severe degeneration and cell loss involve mostly neuronal cells. Thus, it appears that NCL proteins may be critical only for the metabolism of neurons. It is uncertain whether this phenomenon is caused by the specific metabolic requirements of a neuron as a postmitotic cell or results from the properties of NCL proteins per se. This chapter will introduce the reader to the complex pathogenesis of NCLs and will discuss the potential consequences of genetic defects of NCL-associated proteins on the biology of the cell.

II. CLN1. INFANTILE FORM OF NCL: DEFICIENCY OF PALMITOYL-PROTEIN THIOESTERASE 1

The first clinicopathologic descriptions of the infantile form of NCL by Haltia and colleagues (1973) indicated that this particular variant of NCL is the most severe and has the earliest onset of symptoms, the fastest course, and the most prominent destruction of brain tissue among all NCL forms. In fact, postmortem studies demonstrated severe brain atrophy with almost total loss of neurons in the cortical mantle with prominent astrogliosis (Figures 2.1a and 2.1b), and loss of myelin in the white matter. Only some neurons in the hippocampal CA1 and CA4 sectors and basal ganglia as well as numerous neurons in the brainstem are preserved. The retina undergoes severe atrophy. Virtually all the structural elements of the central nervous system show accumulation of the PAS- (Figure 2.1c) and Sudan black B-positive material (Figure 2.1d) revealing autofluorescence when examined under ultraviolet light. Lysosomal storage is visible in other tissues and

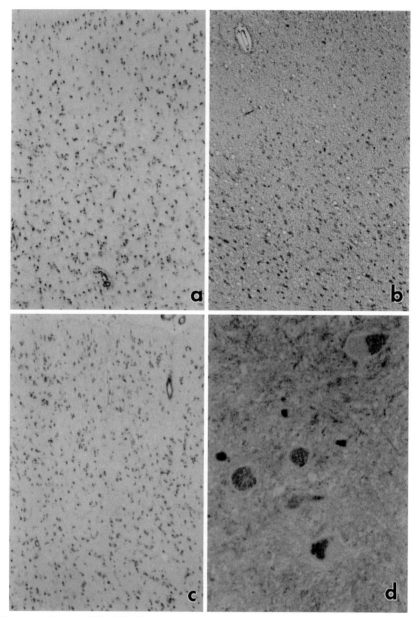

Figure 2.1. Infantile NCL (CLN1). Severe atrophy of the cerebral cortex with total loss of neurons (a) and proliferating glial cells (b) loaded with PAS-positive storage material (c). Preserved neurons of the hypoglossal nucleus accumulating Sudan black-B-positive storage material (d). (a) Cresyl violet, × 100. (b) Polyclonal antibody to glial fibrillary acidic protein (GFAP), × 100. (c) PAS, × 100. (d) Sudan black B, × 400.

organs; however, it is not accompanied by any significant cell loss. The storage material corresponds ultrastructurally to granular osmiophilic deposits (GROD) (Wisniewski et al., 1992; see also Chapter 1) and constitutes approximately 35% lipid and 45% proteins, the latter containing mostly SAPs A and D (Tyynelä et al., 1993).

The CLN1 disease process is associated with mutations of the gene on chromosome 1 that encodes palmitoyl-protein thioesterase 1 (PPT1) (Vesa et al., 1995). To date, 24 different mutations and two polymorphisms in the PPT1 gene have been found. Most PPT1 mutations are predicted to be functionally "null." However, phenotype/genotype correlations showed that some CLN1 subjects have a later onset of clinical signs and symptoms and a more protracted course resembling the late-infantile or juvenile rather than typical infantile form (Das et al., 1998; Mitchison et al., 1998; Wisniewski et al., 1998a; Waliany et al., 2000; for a recent review, see Mole et al., 1999). Because some PPT1 mutants disclose residual enzyme activity (Verkruyse and Hofmann, 1996; Das et al., 1998), it appears that this clinical variability can be explained by the fact that some mutations affect PPT1 function in a more prominent manner than the other mutations. However, the translation of these genetic data into the pathogenesis of CLN1 still remains elusive.

PPT1 is a thioesterase that removes fatty acyl chains, preferentially those that have 14–18 carbons, such as palmitate, myristate, stearate, and oleate, from select cysteine residues in proteins in vitro and is localized to lysosomes (Figure 2.2, see color insert) (Camp and Hofmann, 1993; Camp et al., 1994; Hellsten et al., 1996; Verkruyse and Hofmann 1996; Sleat et al., 1996; see also Chapter 3). PPT1 expression in rat brain is developmentally regulated and stronger in neurons than in glial cells (Suopanki et al., 1999). In mouse brain, neurons, astrocytes, oligodendrocytes and myelin are immunoreactive (Isosomppi et al., 1999). The pattern of PPT1 immunostaining in the normal human infant brain is similar to that in mouse brain, but the immunoreactivity is weaker in the cerebral cortex (Isosomppi et al., 1999).

Studies in cell culture conditions showed that the Arg122Trp mutation in PPT1 that is present in all Finnish CLN1 subjects leads to production of a nonfunctional protein that is degraded in the endoplasmic reticulum (ER) (Hellsten et al., 1996). The cellular fate and biological consequences of other PPT1 mutants remain to be determined. Furthermore, the question of why PPT1 deficiency leads to dramatic brain tissue damage cannot yet be answered. Palmitoylation is a posttranslational modification of proteins leading to the attachment of fatty acids to the side chain of cysteine residues via thioester linkage. Palmitoylation plays a role in the association of polypeptides with membrane or cytoskeleton, prevention of dimerization of proteins, restriction of the lateral mobility of proteins, or receptor recycling. Palmitoylation/depalmitoylation is a reversible and dynamic process, which may regulate the function of numerous proteins of the central nervous system that are implicated in important biological processes such as neuronal transmission, signaling pathways, or protein transport (for recent reviews, see Bizzozero, 1997; Dunphy and Linder, 1998). Thus, potential impairment of these processes by nonfunctional PPT1 could indeed induce a deleterious effect on the functioning of the central nervous system and cause prominent neuronal death. Both transmembrane and soluble proteins can be palmitoylated. However, palmitoylated transmembrane proteins including G-protein-coupled receptors are usually modified either in a juxtamembranous region, near the last transmembrane segment, or at the inner surface of the plasma membrane of cells, whereas PPT1 is localized to lysosomes. Thus, it is reasonable to expect that PPT1 could be implicated in the processes delineated above only if it also is active in other cellular sites than lysosomes. Such a possibility was indeed proposed recently by Isosomppi et al. (1999), based on analysis of PPT1 expression in the mouse brain. However, more direct evidence is needed to validate this intriguing proposal.

Precise pathomechanisms leading to lysosomal storage in CLN1 subjects also are still unclear. Two models explaining the origin of lysosomal storage have been proposed recently. A hypothesis formulated by Bizzozero (1997) is based on previous observations indicating that palmitoylation protects some proteins against lysosomal proteolytic attack. Thus, lack of PPT1 activity and preservation of palmitate groups on proteins could render these particular proteins resistant to lysosomal degradation leading to lysosomal storage. Because physiological substrates for PPT1 are still unknown, verification of this hypothesis is at present difficult. A model proposed by Hofmann and colleagues is based on their finding showing that small lipid thioesters derived from acylated proteins accumulate in lysosomes from CLN1 lymphoblasts as a result of PPT deficiency and defective autophagocytic proteolysis. Thus, it was proposed that accumulation of these compounds could compromise activity of those lysosomal proteases that are sensitive to thiol compounds, leading to more global lysosomal defect than that caused by pure dysfunction of mutated PPT 1 (Lu et al., 1996; Hofmann et al., 1997).

However, as was revealed by Tyynelä et al. (1993), lysosomal storage in CLN1 subjects contains mostly SAPs A and D. This finding raises the question of whether there are any direct or indirect interactions between SAPs and PPT1, or whether their metabolic pathways converge. Accumulation of SAPs represents a very early event of CLN1 pathogenesis because it was evidenced already in diagnostic brain biopsy specimens (Tyynelä et al., 1995). SAPs also were found in amounts varying between 2% and 10% of the molar amount of stored subunit c in other NCL forms (Tyynelä et al., 1995, 1997a), in the storage material in the miniature Schnauzer dog (Palmer et al., 1997b), and congenital ovine NCL (Tyynelä et al., 1997a). Accumulation of SAPs is not restricted to the NCLs. Increased levels of SAPs were found in liver and/or brain tissue from patients with type 1 G_{M1} gangliosidosis, type A Niemann-Pick disease, Tay-Sachs disease,

Sandhoff disease, type 2 G_{M1} gangliosidosis, metachromatic leukodystrophy, and Krabbe disease (Inui and Wenger, 1983). The highest levels of SAPs were found in storage disorders in which the stored lipid(s) bound to SAPs. Furthermore, SAPs are increased in plasma of patients with numerous different lysosomal storage disorders (Chang et al., 2000). Of interest, although in the majority of these disorders all saposins were increased in the plasma, in some of them, including MPS II, only one form of SAPs was elevated.

SAPs A, B, C, and D are small, 8- to 13-kDa glycoproteins that are derived by the proteolytic processing of a 73-kDa precursor protein, prosaposin. Prosaposin is a multifunctional protein encoded in humans by a gene on chromosome 10 (Inui et al., 1985). Proteolytic cleavage of prosaposin to mature saposins through various intermediates occurs in late endosomes and lysosomes and shows cell type-specific differences (Leonova et al., 1996). In vitro studies suggested that cathepsin D may be involved in the maturation of saposins (Hiraiwa et al., 1997). SAPs are lysosomal proteins that activate several lysosomal hydrolases that are engaged in degradation of glycophospholipids. Mutations in SAPs B and C, by themselves, can lead to lysosomal storage disorders. SAP B enhances the cleavage of sulfatide by arylsulfatase A, and its genetic defect leads to a variant of metachromatic leukodystrophy (Holtschmidt et al., 1991). SAP C is a physiological activator of acid β-glucosidase, enabling optimal cleavage of glucosylceramidase, and its deficiency causes a variant form of Gaucher disease (Christomanou et al., 1989).

The mechanisms of action of SAPs are not fully understood (for reviews, see Sandhoff et al., 1995; Vaccaro et al., 1999). It was postulated that SAP B solubilizes and partially extracts specific glycosphingolipids from membranes, making them accessible to the active site of arylsulfatase A, or that SAP C stimulates glucosylceramide degradation by interacting with phospholipid membranes and favoring the glucosylceramidase localization on lipid surfaces. Furthermore, some SAPs show high affinity for phospholipids and are involved in binding and intervesicular transport of glycosphingolipids. Thus, although lysosomal accumulation of SAPs may represent an epiphenomenon of no or minor significance, the role of potential dysregulation of SAP metabolism in the pathogenesis of CLN1 cannot be excluded.

Of interest, prosaposin also is secreted and then taken up by the cells via mannose 6-phosphate-dependent; mannose-dependent, and low-density lipoprotein receptor-related protein-dependent routes (Hiesberger et al., 1998). However, secreted prosaposin also may bind to a specific cell surface receptor and function as a signaling molecule with neurotrophic and antiapoptotic activity (O'Brien et al., 1994; Hiraiwa et al., 1997; Tsuboi et al., 1998). The neurotrophic activity encompasses a region spanning residues 22–31 in the N-terminal part of the saposin C domain (Qi et al., 1999). Prosaposin stimulates neurite outgrowth and prevents death of a variety of neuronal cells. In addition, it prolongs survival of Schwann cells and oligodendrocytes, acting as myelinotrophic factors (Hiraiwa *et al.*, 1997). Thus, diminished neurotrophic and antiapoptotic action of prosaposin could potentially contribute significantly to the massive neuronal death observed in CLN1 subjects.

It cannot be excluded at present that saposins interact in some way with PPT1. All SAPs have a similar location of six cysteins, and their disulfide arrangement has been found to be essential for their action (O'Brien and Kishimoto, 1991; Vaccaro *et al.*, 1995). SAPs could potentially represent natural substrates for PPT1. Therefore, lack of putative depalmitoylation of SAPs caused by PPT1 deficiency could affect proper function of these activator proteins. PPT1 deficiency also could be involved indirectly in the metabolism of SAPs, i.e., by changing the lipid content of the lysosomal milieu. These complex aspects of the CLN1 disease process require further experimental studies.

III. CLN2. THE CLASSIC LATE-INFANTILE NCL: DEFICIENCY OF TRIPEPTIDYL-PEPTIDASE I

Severe brain atrophy with massive neuronal loss in the cerebral and cerebellar cortices represent a feature of classical late-infantile NCL brain (Figures 2.3a–2.3d). Less severe neuronal loss is found in subcortical gray matter areas, and the least in brainstem nuclei. Diffuse proliferation of macrophages most probably of microglial origin (Brück and Goebel, 1998) and astrogliosis, both most prominent in the cerebral cortex, represent a constant structural abnormality. Retinae show atrophy. Autofluorescent, PAS- and Sudan black B-positive lysosomal storage material that is present in all tissues and organs is visualized ultrastructurally as curvilinear profiles (Wisniewski *et al.*, 1992, see also Chapter 1). However, the intensity of neuronal damage varies, and neurons with relatively well preserved structure (Figure 2.4a), as well as neurons severely damaged and overloaded with lysosomal storage coexist in the same brain areas (Figure 2.4b).

A main component of the lysosomal storage material in CLN2 subjects is subunit c of mitochondrial ATP synthase (Figures 2.5a and 2.5b), which constitutes around 85% of the protein content of stored cytosomes (Palmer *et al.*, 1989, 1992; Fearnley *et al.*, 1990; Kominami *et al.*, 1992). Subunit c of mitochondrial ATP synthase also accumulates in other NCL forms, except for the CLN1 variant (Hall *et al.*, 1991a; Elleder *et al.*, 1997; Tyynelä *et al.*, 1997b; Haltia *et al.*, 1999); in some canine, bovine, and ovine forms of NCLs (for review, see Palmer *et al.*, 1997a); and in *mnd/mnd* (for motor neuron degeneration) mice (Pardo *et al.*, 1994; Faust *et al.*, 1994). In addition, accumulation of subunit c was detected in lysosomal storage diseases other than NCLs (Kida *et al.*, 1993; Elleder *et al.*, 1997). However, even if accumulation of subunit c was conspicuous and widespread in the neuronal population in some of these conditions, such as in mucopolysaccharidoses I, II, IIIA, subunit c staining was generally absent in non-neuronal

2. Cellular Pathology and Pathogenic Aspects of NCLs 43

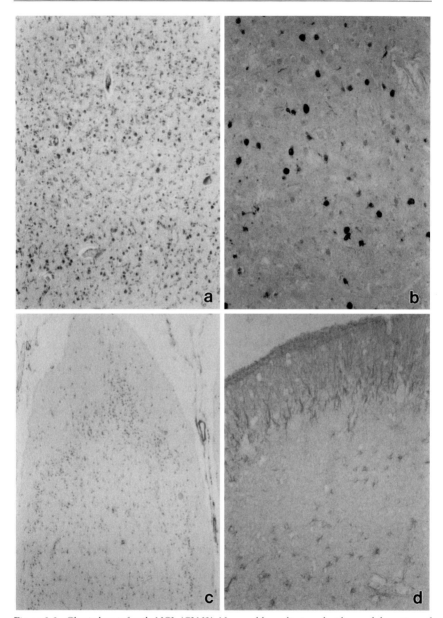

Figure 2.3. Classic late-infantile NCL (CLN2). Neuronal loss, gliosis and widespread deposition of PAS-positive material in neurons, glial cells and blood vessel walls in the neocortex (a). Numerous macrophages in the cerebral cortex (b). Neuronal loss in the granular and pyramidal layers of the cerebellar cortex (c) with gliosis (d). (a) PAS, × 100. (b) Anti-ferritin immunostaining, × 200. (c) PAS, × 100. (d) Anti-GFAP immunostaining, × 200.

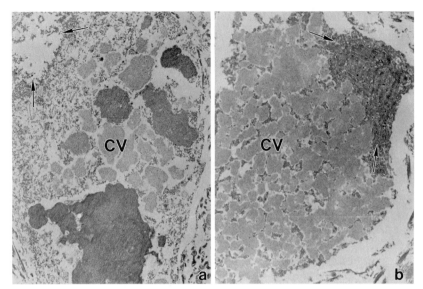

Figure 2.4. Classic late-infantile NCL (CLN2). Substantia nigra neuron with relatively well preserved structure (a), and severely damaged neuron in the same brain area (b). Arrows point to the nuclei. CV indicates lysosomal storage material. Electron microscopy. Postmortem tissue. Orig. magn. × 5000.

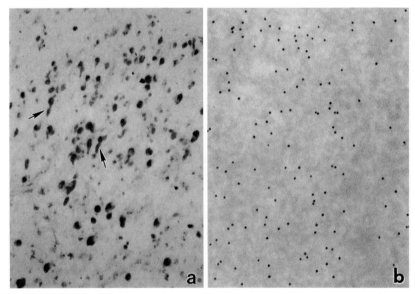

Figure 2.5. Classic late-infantile NCL (CLN2). (a) Accumulation of subunit c of mitochondrial ATP synthase in neurons and glia in the cerebral cortex. Arrows indicate the severely distended proximal part of the axon (meganeurites). (b) Immunoelectron microscopically, subunit c of mitochondrial ATP synthase is associated with curvilinear profiles. Immunogold method with 10 nm gold particles. (a) × 200; (b) × 40,000.

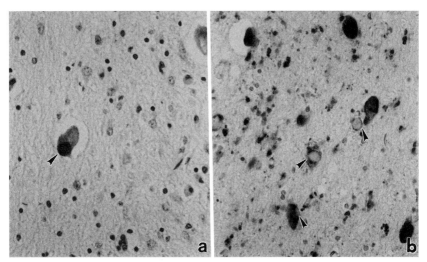

Figure 2.6. Classic late-infantile NCL (CLN2). Neuronal inclusion bodies (arrowheads) easily identifiable in Klüver-Barrera staining (a) are not immunoreactive to subunit c of mitochondrial ATP synthase (b). (a), (b), × 400.

cells of the brain and visceral organs (Elleder et al., 1997). This is in sharp contrast to CLN2, where deposition of subunit c-immunopositive material is observed in many tissues and organs. Furthermore, in CLN2 and some CLN3 subjects, subunit c deposition in kidney is so prominent that it can be detected in urine samples, representing an easy screening test for these disorders (Wisniewski et al., 1994, 1995).

Neurons in some brain regions of CLN2 subjects, such as the nucleus dentatus, nucleus subthalamicus, or substantia nigra, contain spheroidal inclusions that are immunoreactive neither to subunit c nor to selected lectins, even though the storage material surrounding them shows the staining (Figure 2.6) (Kida et al., 1995). These inclusions can be especially numerous in some CLN2 cases (Wisniewski et al., 1993) and most probably correspond to the so-called myoclonus bodies described originally by Seitelberger and colleagues (1967). Elleder and Tyynelä (1998) reported recently the presence of SAPs-negative storage bodies in CLN1, CLN2, and CLN6 subjects and subunit c-negative deposits also in CLN6 subjects. The origin of these inclusions is at present unclear. It was proposed that the lysosomal storage material undergoes gradual modifications, which may lead to loss of immunodetectable epitopes of both subunit c and SAPs. It is also conceivable that these inclusions may contain other components of the storage material than subunit c or SAPs, such as lipid thioesters or small polypeptides (see below).

Subunit c is an intrinsic membrane component of the ATP synthase complex located in the inner mitochondrial membrane, where it is involved in rotary motion of the ATP synthase complex and proton transport through the membrane (Fillingame, 1997). The 75-amino acid sequence and 42-dalton modification of subunit c isolated from the storage material do not differ from normal subunit c (Fearnley et al., 1990; Palmer et al., 1992; Buzy et al., 1996). None of the three nuclear genes, P1, P2, and P3, coding for subunit c located on chromosomes 17, 12, and 2, respectively (Dyer and Walker, 1993; Yan et al., 1994), could be associated with the NCL disease process (Medd et al., 1993; Ezaki et al., 1995). Thus, it was proposed that accumulation of subunit c of mitochondrial ATP synthase is caused by decreased degradation of this highly hydrophobic proteolipid (Palmer et al., 1989, 1992; Kominami et al., 1992; Ezaki et al., 1995).

Most of the experimental studies aimed at characterizing the role of subunit c accumulation for CLN2 pathogenesis were done by using primary fibroblast cultures. However, accumulation of subunit c in vitro in these particular types of cells was controversial. Both our (Kida et al., 1993) and others studies (Lake and Rowan, 1997) disclosed neither significant subunit c storage nor the presence of curvilinear profiles in cultured fibroblasts from CLN2 subjects. However, accumulation of subunit c in cultured fibroblasts was reported by others (Tanner and Dice, 1995; Tanner et al., 1997; Kominami et al., 1992; Ezaki et al., 1995, 1997, 1999). In a series of publications, Ezaki and colleagues (Kominami et al., 1992, Ezaki et al., 1995, 1996, 1997) presented data showing that although the majority of subunit c from control fibroblasts was found in fractionated mitochondria, around 50–70% of subunit c from patients' cells was recovered in lysosomal fractions. There also was delayed intramitochondrial and lysosomal degradation of labeled subunit c in fibroblasts from CLN2 patients. Furthermore, only isolated lysosomal fractions from control but not CLN2 fibroblasts were able to degrade subunit c either from control or CLN2 cells. Nevertheless, pulse chase experiments presented by these researchers disclosed unequivocally that fibroblasts from CLN2 patients did in fact degrade subunit c, but less efficiently than control cells. This suggests that the activity of the enzyme involved in subunit c degradation is impaired but not abolished totally in the cells studied. It cannot be excluded that more than one protease is involved in subunit c degradation in vivo as well, assuming that CLN2 protein is implicated directly in subunit c proteolysis.

In 1997, by using a proteomic approach, Sleat and colleagues identified the protein defective in CLN2 subjects (Sleat et al., 1997). The significant similarity of this protein to a specific group of bacterial proteases suggested that this protein is pepstatin-insensitive carboxyl endopeptidase. However, Rawlings and Barrett (1999) disclosed that CLN2-deficient protein corresponds in fact to lysosomal tripeptidyl-peptidase I (TPP I), an enzyme purified recently to homogeneity from the rat spleen by Vines and Warburton (1998). TPP I is an aminopeptidase that releases different tripeptides from a free N-terminus, with preference to hydrophobic amino acids in the P_1-position (Page et al., 1993), and also shows endopeptidase activity *in vitro* (Ezaki *et al.*, 2000). With the use of immunocytochemistry, TPP I was localized in the brain to both neurons and glia (Oka *et al.*, 1998). An acidic pH optimum together with the presence of mannose 6-phosphate modification suggested that CLN2p acts within lysosomal compartments (Sleat *et al.*, 1997, 1999; Sohar *et al.*, 1999). The lysosomal localization of TPP I was indeed documented afterwards by immunofluorescence and subcellular fractionation (Ezaki *et al.*, 1999).

Efficient protein degradation in lysosomes depends on concerted action of endopeptidases and exopeptidases, which provide dipeptides and free amino acids exported to the cytoplasm and further used according to the metabolic needs of the cell. Thus, TPP I could provide tripeptides for subsequent action of dipeptidyl peptidases and dipeptidases, and its dysfunction could lead to the accumulation of undigested polypeptides in the lysosomal lumen. Of interest, accumulation of low-molecular-weight compounds was observed previously in the storage material (Palmer *et al.*, 1993), but approaches to provide their closer biochemical characterization were unsuccessful.

It appears that most of 26 mutations identified until present in the TPP I gene cause loss of TPP I activity (Sleat *et al.*, 1999; Sohar *et al.*, 1999; Junaid *et al.*, 1999; Vines and Warburton, 1999; for a recent review, see Mole *et al.*, 1999). The cellular fate of mutated TPP I has not yet been determined. However, TPP I is undetectable in fibroblasts and brain and liver tissue from CLN2 subjects (Oka *et al.*, 1998; Ezaki *et al.*, 1999) (Figure 2.7), which suggests that mutated TPP I may be degraded soon after its synthesis. Although TPP I cleaves in vitro peptide hormones, such as growth hormone (Doebber *et al.*, 1978), angiotensin II, and glucagon (Vines and Warburton, 1999), or substance P (Junaid *et al.*, 2000), the natural substrates for TPP I have to be determined. Of interest, Junaid and colleagues (2000) demonstrated recently that TPP I also is involved in hydrolysis of synthetic amyloid-β peptides 1-42 and 1-28. Accumulation of amyloid-β peptide in the brain tissue of CLN2 subjects was suggested by our previous studies (Kitaguchi *et al.*, 1990; Wisniewski *et al.*, 1990a, 1990b).

Elleder and colleagues (1995) showed that various cell types tested under cell culture conditions were not able to degrade subunit c isolated from the storage material from affected animals. However, Ezaki and colleagues (1999) showed that immunodepletion of TTP I from a lysosomal fraction prepared from control cells led to diminished degradation of mitochondrial subunit c from control cells. In agreement with this observation, Junaid and colleagues (2000) found that TPP I isolated from the bovine brain was able to cleave off a tripeptide from a synthetic amino-terminal hexapeptide or carboxy-terminal pentapeptide from the subunit c sequence. Both these observations suggested strongly that subunit c of mitochondrial ATP synthase represents a substrate for TPP I. However, subunit c isolated from the storage material from CLN2 subjects was not degraded in vitro

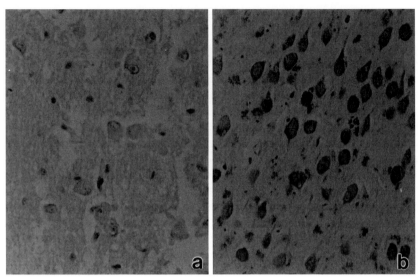

Figure 2.7. Lack of tripeptidyl-peptidase I immunoreactivity in the cerebral cortex of CLN2 subject homozygous to Arg208STOP nonsense mutation (a) and strong immunostaining in the cerebral cortex of CLN3 subject (b). Monoclonal antibody 8C4, × 200.

by purified bovine TPP I (Junaid et al., 2000). Thus, it appears that further studies are needed to ascertain that subunit c accumulation in CLN2 subjects is caused directly by TPP I deficiency.

Whether and how subunit c accumulation is associated with progressive neuronal degeneration and death remains unknown. Recently, it was reported that subunit c also is present in the plasma membrane, where it acts as a cation pore (McGeoch and Guidotti, 1997). Thus, it was proposed that increased subunit c content in the neurons could render these cells more susceptible to depolarization and increased excitability, leading to their death (McGeoch and Palmer, 1999). Of interest, in some NCL forms and in the *mnd/mnd* mice, also small amounts of subunit c of vacuolar ATPase have been identified (Faust et al., 1994; Palmer et al., 1997a). Subunits c of vacuolar ATPase reveal a homology to subunit c of mitochondrial ATPase and are derived by gene duplication and fusion from the mitochondrial ATPase c subunit (for review, see Forgac, 1998). Vacuolar ATPase is associated with membranes of intracellular acidic vesicular compartments, where it maintains pH gradient across the membrane. However, vacuolar ATPase also is expressed at the plasma membrane of some types of cells. Thus, it cannot be excluded at present that subunit c of mitochondrial ATP synthase and subunit c of vacuolar ATPase deposited in the storage material originate at least partially from the plasma membrane. If this scenario is true, both endocytosis and macroautophagy could be involved in lysosomal transport of subunits c.

It is also conceivable that the cell damage observed in CLN2 subjects may result from defective degradation and lysosomal storage of small peptides as a direct consequence of TPP I dysfunction (Vines and Warburton, 1999). Another interesting pathogenic hypothesis was presented recently by Das and colleagues (1999). These researchers disclosed deficiency of mitochondrial ATP synthase in various NCL forms as well as in the sheep model of NCL. Thus, they proposed that deficient ATP supply might represent a common denominator of neuronal degeneration in NCLs affecting especially those neurons that have high-energy requirements. Also, this hypothesis needs more experimental data to be confirmed.

IV. CLN3. JUVENILE FORM OF NCL: GENETIC DEFECT OF LYSOSOMAL MEMBRANE PROTEIN

Similar to CLN1 and CLN2, the CLN3 disease process affects most severely the brain tissue. However, brain atrophy and neuronal loss (Figures 2.8a and 2.8b) are distinctly less prominent in subjects with the juvenile variant of NCLs than in CLN1 and CLN2 individuals. In the cerebral cortex, neuronal loss predominates in layer V (Braak and Goebel, 1979). The cerebellar cortex shows widespread, moderate loss of granule cells, whereas the pyramidal layer is less affected. Neuronal loss is rather mild in subcortical nuclei and brainstem. Gliosis, manifested mostly by proliferation of astrocytes and, to a lesser degree, of microglia,

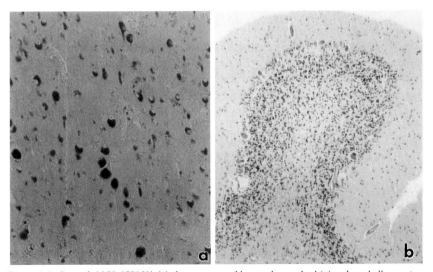

Figure 2.8. Juvenile NCL (CLN3). Moderate neuronal loss in the cerebral (a) and cerebellar cortices (b). (a) Sudan black B, × 100. (b) PAS, × 100.

is distinctly less severe than in CLN1 and CLN2 subjects. Retinae show severe atrophy with gliosis. Autofluorescent, Sudan black B- and PAS-positive storage material is widespread in the brain tissue, often being especially abundant in pyramidal neurons of hippocampal sectors CA2–CA4. Lysosomal storage is not confined to the brain but also is present in various tissues and organs. Ultrastructurally, the storage material is composed of fingerprint profiles. Vacuolated lymphocytes are typical for this form of NCLs (Wisniewski et al., 1992, see also Chapter 1).

Subunit c of mitochondrial ATP synthase constitutes around 20% of the protein content of the storage material (Palmer et al., 1992). Other identified constituents of the lysosomal storage material include lipids, phosphorylated dolichols (for review, see Hall et al., 1991b), small amounts of SAPs (Tyynelä et al., 1995), and amyloid-β peptide (Wisniewski et al., 1990a, 1990b). Of interest, the level of numerous mannose 6-phosphate-modified lysosomal enzymes is increased significantly—even up to sevenfold—in brain tissue of JNCL subjects (Sleat et al., 1998).

Although the CLN3 gene was identified in 1995 (International Batten Disease Consortium, 1995), our knowledge of the role of the CLN3 protein (CLN3p) in the cell biology is very limited. The major problem encountered by researchers working on unraveling the function of CLN3p was posed by difficulties in raising antibodies recognizing endogenous CLN3p, which was caused most probably by specific structural and functional properties of CLN3p. As a consequence, most of the studies were carried out using mammalian expression systems.

Structural modeling of CLN3p has led to the suggestion that CLN3p might be a mitochondrial protein (Janes et al., 1996). Indeed, Katz and colleagues (1997) reported the mitochondrial localization of CLN3p in Müller cells and some inner neurons of mouse retina by using polyclonal antibodies to CLN3 protein. On Western blot, this antibody recognized a 50-kDa band. However, the specificity of this antibody was not confirmed by using recombinant CLN3p and could not be verified by Western blot analysis of disease and normal human material, because it did not react with human tissue. The immunofluorescence method used in another recent study by Järvelä and colleagues (1998) showed that CLN3p expressed in HeLa cells and detected by a polyclonal antibody raised in rabbit colocalizes with a lysosomal-associated membrane protein, lamp-1, and thus is targeted to lysosomal compartments.

Our studies performed by using a different methodology confirmed that CLN3p either expressed in fusion with the green fluorescent protein or tagged with FLAG epitope is a lysosomal membrane protein (Kida et al., 1999; Golabek et al., 1999, 2000) (Figures 2.9a–2.9c). Lysosomal localization of yeast homologue of CLN3p, BTN1, also was disclosed (Croopnick et al., 1998; Pearce et al., 1999a). However, it also was suggested that a substantial portion of overexpressed CLN3p is associated with the Golgi apparatus (Kremmidiotis et al., 1999).

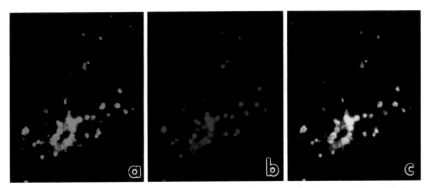

Figure 2.2. Lysosomal localization of palmitoyl-protein thioesterase 1 in human embryonal kidney cell (HEK293). Subcellular distribution of lamp-1 (lysosomal marker) (a, green) and PPT1 (b, red) overlap on a composite image (c, yellow) generated by laser-scanning confocal microscope. × 600.

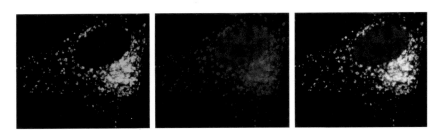

Figure 2.10. Lysosomal localization of CLN3p in human neuroblastoma cell. CLN3p fused to the green fluorescent protein (a, green) shows similar subcellular distribution to that of lamp-1, lysosomal marker (b, red) on a composite image (c, yellow), generated by laser-scanning confocal microscopy. × 600.

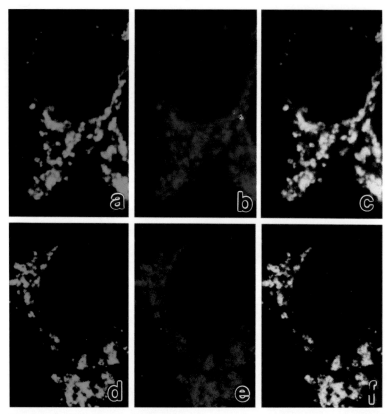

Figure 2.11. Both wild-type CLN3p fused to the green fluorescent protein (a, green) and CLN3p bearing Arg334Cys mutation (d, green) are targeted to lysosomal compartments, as visualized on composite images (c, f, yellow) generated by laser-scanning confocal microscopy of HEK293 cells immunostained with antibodies to lamp-1 (lysosomal marker) (b, e, red). × 600.

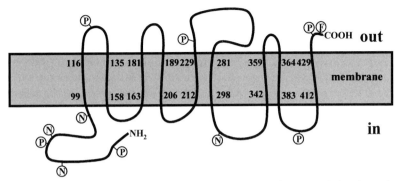

Figure 2.9. Predicted structure of CLN3p. The schematic structure of CLN3p including the predicted N-glycosylation (N) and phosphorylation (P) sites was drawn on the basis of TMpred, SOSUI, and MotifFinder algorithms. For the clarity of the picture, the predicted posttranslational lipid modifications were omitted, except for farnesylation motif (F) at the C-terminus.

Analysis of the predicted amino acid sequence of CLN3p suggested that CLN3p belongs to type-IIIb membrane proteins (von Heijne and Gavel, 1988) and that, according to hydropathy calculation, it may contain six putative transmembrane domains (Janes et al., 1996). Nevertheless, modeling with several other hydrophobicity algorithms suggests that CLN3p may span up to nine transmembrane regions (Figure 2.10, see color insert). The specific amino acid composition dictates that CLN3p is highly hydrophobic. In addition, its molecule contains numerous structural motifs that can be engaged in extensive co- and posttranslational modifications. According to the analysis performed by The International Batten Disease Consortium (1995), CLN3p has four potential N-linked glycosylation sites at residues 49, 71, 85, and 310; two putative O-glycosylation sites at residues 80 and 256; two potential glycosaminoglycan sites at residues 162 and 186; six protein kinase C phosphorylation sites; two cAMP- and cGMP-dependent protein kinase phosphorylation sites; eight casein kinase II phosphorylation sites; and 12 N-myristoylation sites and palmitoylation sites.

Our studies showed that CLN3p, like other lysosomal membrane proteins (Hunziker and Geuze, 1996) is glycosylated extensively and that the pattern of glycosylation of CLN3p shows cell-type-specific differences (Golabek et al., 1999). According to our data, CLN3p also is phosphorylated (Michalewski et al., 1998), most probably processed proteolytically (Golabek et al., 2000), and farnesylated (Kaczmarski et al., 1999). The biological significance of these co- and posttranslational modifications is at present unclear. However, characterization of co- and posttranslational modifications of CLN3p is important to understand the function of CLN3p, given that posttranslational modifications may significantly determine the properties and/or regulate the activity of proteins.

The most common mutation in the CLN3 gene is a 1.02-kb deletion found in 85% of the Batten disease chromosomes examined (Munroe et al., 1997). This deletion, affecting either one or both alleles, produces a frameshift and generates a predicted translation product of 181 amino acids (153 residues of the protein, followed by 28 novel amino acids before the stop codon), which is retained and most probably degraded in the endoplasmic reticulum (Järvelä et al., 1999).

Apart from this common deletion, 24 other mutations, including deletions, insertions, and point mutations, have been discovered (Munroe et al., 1997; Wisniewski et al., 1998b; Zhong et al., 1998; Lauronen et al., 1999; for a recent review, see Mole et al., 1999). It was suggested that these mutations either result in severely truncated CLN3p or affect its structure/conformation. Phenotype/genotype correlation suggests that there are structural domains in CLN3p that are critical for CLN3 protein function, and that, even if affected by point mutations, induce a severe deleterious effect in vulnerable tissue. It is still unknown which particular properties of CLN3p are changed by point mutations. Mutations present in transmembrane domains can change the normal targeting of lysosomal membrane proteins by altering their membrane incorporation, as was documented for lysosomal membrane protein, lamp-1 (Roher et al., 1996; Wimer Mackin and Granger, 1996). Mutated proteins also can be misfolded and degraded rapidly in the endoplasmic reticulum (Klausner and Sitia, 1990), which might explain why some mutated proteins may be virtually undetectable in the tissues of affected subjects. Our unpublished data suggest that both CLN3-101 mutant (Leu101Pro) and CLN3-334 mutant (Arg334Cys) are targeted to lysosomal compartments (Figures 2.11d–2.11f, see color insert), thus, similar to wild-type CLN3p (Figures 2.11a–2.11c, see color insert).

Lysosomal-associated membrane proteins such as lamp-1, lamp-2, limp-I, or limp-II are targeted to lysosomes either through late endosomes or through a secretory pathway and endocytosis from the plasma membrane. These proteins utilize two different structural lysosomal sorting motifs located in their C-terminal tails: either tyrosine-based or di-leucine-based signals (Fukuda, 1994; Sandoval et al., 1994; Marks et al., 1996). However, our data suggest that a di-leucine motif present in the C-terminal fragment of CLN3p is not involved in its lysosomal targeting in vitro (Kida et al., 1999). In this respect, CLN3p recalls sialin and other transporters of the ACS family, which do not possess a known targeting signal (Verheijen et al., 2000).

There is no significant sequence homology between CLN3p and any other protein known in mammals. However, the yeast homolog in *Saccharomyces cerevisiae*, *btn1*, encodes a nonessential protein that shows 39% identity and 59% similarity to CLN3. Btn1-deletion yeast strains show pH-dependent resistance to D-(−)-threo-2-amino-1-[p-nitrophenyl]-1,2-propanediol (ANP), a phenotype that can be complemented by expression of human CLN3 (Pearce and Sherman,

1998). Furthermore, *btn1-*Δ yeast strains have an abnormally acidic vacuolar pH in the early phases of growth, which can be reversed by the presence of chloroquine in the growing medium (Pearce and Sherman, 1998; Pearce et al., 1999a, 1999b). These data indicate that BTN1 is involved in the control of vacuolar pH in yeast (for details, see Chapter 11).

To study the function of CLN3p, we used human embryonal kidney cells (HEK-293) stably expressing wild-type and mutant CLN3p as well as cells stably transfected with antisense-oriented CLN3 cDNA under the control of the ecdysone-inducible promoter (Golabek et al., 2000). Our data indicate that wild-type CLN3p increases lysosomal pH in HEK-293 cells, whereas inhibition of CLN3p synthesis by the antisense approach acidifies lysosomal compartments. In addition, we found that this biological activity of wild-type CLN3p affects intracellular processing of amyloid-β protein precursor and cathepsin D, proteins whose metabolism is influenced by pH of intracellular acid compartments. We also found that mutant CLN3p (Arg334Cys), which is associated with the classical juvenile NCL phenotype in compound heterozygotes, is devoid of the biological activity of wild-type CLN3p. Of interest, our previous studies suggested accumulation of amyloid-β protein in the brain tissue of CLN3 subjects (Kitaguchi et al., 1990; Wisniewski et al., 1990a, 1990b). Thus, both our data and observations of Pearce and colleagues suggest that the pathogenesis of juvenile NCL is associated with altered acidification of lysosomal compartments.

The acidic pH of lysosomes is generated as a result of the activity of vacuolar H^+-ATPase (V-ATPase), a major electrogenic proton pump in eukaryotic cells (for review, see Forgac et al., 1998). The V-ATPase pumps protons from the cytosol to the vacuoles by using the energy released by ATP hydrolysis. Acidification of vacuolar compartments plays an important role in various cellular processes such as membrane trafficking, formation of endosomal carrier vesicles, storage of metabolites, or dissociation of mannose 6-phosphate receptors from their ligands. Maintaining the acidic pH of the lysosomal milieu may be critical for the activity of numerous lysosomal enzymes, taking into account that most lysosomal enzymes have acidic pH optima. Derangement of the proper pH of the lysosomal lumen may affect predominantly the activity of those lysosomal enzymes that are especially sensitive to pH changes. Maintenance of the transmembrane pH gradient by a vacuolar H^+-ATPase also may determine the normal posttranslational proteolysis of some proteins (Beers, 1996). In addition, a rise in lysosomal pH induced by lysosomotropic amines, $H(+)$-ionophores, or inhibition of vacuolar type $H(+)$-ATPase, may increase lysosomal secretion (Braulke et al., 1987).

However, some lysosomal enzymes such as cysteine endopeptidases are active over a broad range of pH 4–8 (for review, see Mason, 1995). Furthermore, other enzymes such as sialic acid-specific O-acetylesterase (Butor et al., 1995) or PPT1, associated with infantile NCL (Verkruyse and Hofman, 1996), have neutral pH optima. Thus, it was suggested that intralysosomal pH can fluctuate, facilitating efficient hydrolytic activity of enzymes with a wide range of pH optima (Butor et al., 1995). We propose that alteration of this process by CLN3 protein deficiency and increased acidification of lysosomes could impair the activity of enzymes with pH optima closer to neutral, leading to lysosomal storage. Further studies are needed to verify this hypothesis.

The cause of neuronal death in CLN3 subjects is not known. Boustany and colleagues presented data suggesting that neurons and photoreceptors die in the mechanism of apoptosis in late-infantile and juvenile forms of NCL as well as in canine and ovine models of NCL (Lane et al., 1996). Further studies of these researchers implied that CLN3p exerts a neurotrophic and anti-apoptotic effect in vitro; thus, lack of functional CLN3p in subjects with the juvenile form of NCL could trigger apoptotic signaling pathways (Puranam et al., 1999). Increased rate of apoptosis also was observed in peripheral blood mononuclear cells from NCL patients (Kieseier et al., 1997). Recently developed CLN3 knockout mice (Katz et al., 1999; Mitchison et al., 1999) will be helpful in further elaborating this issue and gaining better insight into the complex pathogenesis of CLN3 disease process.

V. OTHER NCL FORMS WITH KNOWN GENETIC DEFECT

Our knowledge of the pathogenesis of the forms of NCL other than CLN1, CLN2, and CLN3 is even more limited. Two of these NCL variants, CLN5 and CLN8, are encountered almost exclusively in the Finnish population and were assigned previously to chromosomes 13 and 8, respectively (Savukoski et al., 1994; Tahvanainen et al., 1994). Only recently, genes encoding proteins associated with the Finnish variant of late-infantile NCL (CLN5) (Savukoski et al., 1998) and Northern epilepsy (CLN8) (Ranta et al., 1999) have been isolated and characterized. The subcellular localization of the proteins encoded by these genes as well as their function are unknown at present. However, analysis of the predicted amino acid sequences of both these proteins suggest that they have transmembrane topology (Figure 2.12) and may undergo significant co- and posttranslational modifications—thus, similar to CLN3p.

The clinical course, pathology, and biochemistry of the CLN5 disease process recalls that observed in classical late-infantile NCL. Major differences include a later onset of the clinical signs, slower progression, less severe brain atrophy, and variable ultrastructure of lysosomal inclusions visualized as fingerprint and curvilinear profiles as well as rectilinear complexes (for a recent review, see Santavuori et al., 1999; see also Chapter 6). Similarly to CLN2, the storage material contains mostly subunit c of mitochondrial ATP synthase and small amounts of SAPs A and D (Figures 2.13a and 2.13b) (Tyynelä et al., 1997b).

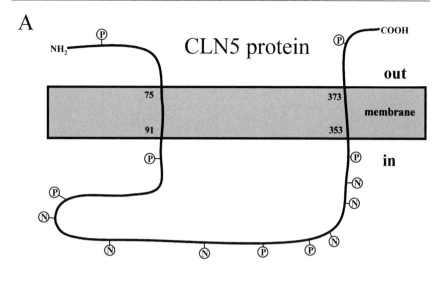

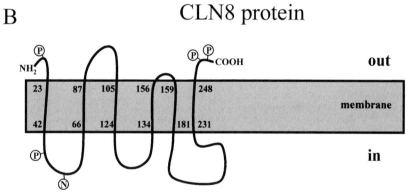

Figure 2.12. Predicted structures of CLN5p (A) and CLN8p (B). The schematic structures of CLN5p and CLN8p, including the predicted N-glycosylation (N) and phosphorylation (P) sites, were drawn on the basis of TMpred, SOSUI, and MotifFinder algorithms. For the clarity of the picture, the predicted posttranslational lipid modifications were omitted.

Northern epilepsy (progressive epilepsy with mental retardation), although known in the Finnish population for many years, has only recently been added to the group of NCLs and its gene assigned as CLN8. Clinical course, symptomatology, and pattern of brain tissue damage in CLN8 subjects differ from those in all other forms of NCL. However, the storage material, similar to other NCL forms, shows immunoreactivity to subunit c of mitochondrial ATP synthase and SAPs, as well as to amyloid-β peptide (for a recent review, see Haltia et al., 1999; see also Chapter 6).

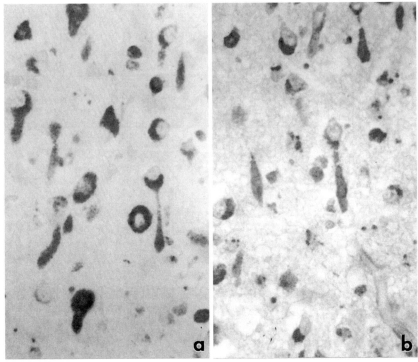

Figure 2.13. Immunoreactivity to subunit c of mitochondrial ATP synthase (a) and to saposin A and D (b) in the frontal cortex of CLN5 subject. From the collection of Dr. J. Tyynelä. (a), (b) × 450.

Of interest, the gene lesion in the *mnd/mnd* mice that exhibit some clinicopathological signs of NCL including accumulation of subunit c of mitochondrial ATP synthase, subunit c of vacuolar ATPase, and dolichol-derived oligosaccharides (Pardo *et al.*, 1994, Faust *et al.*, 1994) is syntenic to CLN8 in humans (Ranta *et al.*, 1999). The availability of a mouse model may be helpful to understand the pathogenesis of CLN8.

VI. NCL FORMS WITHOUT IDENTIFIED GENETIC DEFECTS

Variant late-infantile/early-juvenile NCL, assigned to CLN6 and mapped to chromosome 15 (Sharp *et al.*, 1999), recalls both the classical late-infantile and juvenile NCL forms. The clinical course and the pattern of brain damage are similar to the late-infantile NCL. However, the severity of neuronal loss in the cerebral cortex is less prominent. The ultrastructure of the storage material

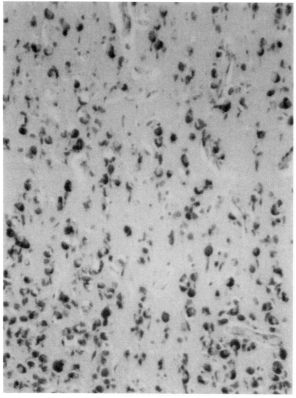

Figure 2.14. Accumulation of subunit c of mitochondrial ATP synthase in neurons and glia of the frontal cortex of CLN6 subject. × 200.

includes mostly fingerprint profiles and rectilinear complexes. In contrast to CLN3, lymphocytes are not vacuolated (for a recent review, see Williams et al., 1999a). The storage material is immunopositive to subunit c of mitochondrial ATP synthase (Figure 2.14) (Elleder et al., 1997).

Recently, Palmer and colleagues developed a cell culture model of fetal neurons from sheep affected with ceroid lipofuscinosis (Hughes et al., 1999; Kay et al., 1999). These cells accumulate significant amounts of subunit c of mitochondrial ATP synthase in vitro. The genetic lesion in these animals is syntenic to the human CLN6 form (Broom et al., 1998). Thus, application of this new model to evaluate molecular mechanisms underlying subunit c storage and its potential influence on cell viability could provide new light on the pathogenesis of this form of NCL.

Only very scanty data are available on the Turkish variant late-infantile NCL, assigned to CLN7. From the clinicopathological point of view, the

CLN7 disease process recalls both the late-infantile and early-juvenile cases. Fingerprint profiles, curvilinear profiles, and rectilinear complexes were reported in the lysosomal storage (for a recent review, see Williams *et al.*, 1999b).

The adult form, CLN4 (Kufs' disease), is a very rare form of NCL and differs in several aspects from the childhood forms (Berkovic *et al.*, 1988). The CLN4 disease process may be manifested by two major phenotypes: type A, with predominance of dementia and behavioral problems; and type B, with progressive myoclonus epilepsy with dementia. In contrast to the childhood forms of NCL, subjects with Kufs' disease have intact vision. Pigmentary retinal degeneration is absent. Inheritance, in contrast to other NCL forms, may be either autosomal recessive (more common) or autosomal dominant (rare). Storage material predominates in the brain tissue. Ultrastructurally, granular, fingerprint, and curvilinear profiles were reported (for details, see Berkovic *et al.*, 1988; Goebel and Braak, 1989; Martin *et al.*, 1999). In the storage material, subunit c of mitochondrial ATPase (Hall *et al.*, 1991a), SAPs A and D (Tyynelä *et al.*, 1997), and amyloid-β peptide (Figure 2.15) (Wisniewski *et al.*, 1990a) were found.

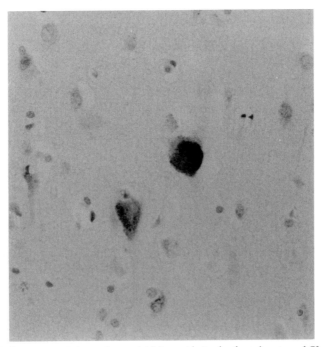

Figure 2.15. Immunoreactivity to the amyloid-β peptide in the frontal cortex of CLN4 subject. Immunostaining with monoclonal antibody 4G8 recognizing amino acids 17–24 of amyloid-β peptide. × 200.

Brain atrophy is usually prominent in the frontoparietal areas and cerebellum. Neuronal loss is the most pronounced in layers II and III of the cerebral cortex and the pyramidal layer of the cerebellum. Of interest, similar to juvenile NCL, among the hippocampal sectors, CA1sector is less affected. The morphology of cells accumulating lipopigment is similar to that observed in childhood forms and includes severe distention of the cell somata and the proximal segment of axons, forming meganeurites. Of interest, in spite of preservation of vision and lack of evident damage to the retina, storage material also was reported in the retinal ganglion cells (Martin et al., 1987).

The gene locus of CLN4 has not yet been identified. By using mannose 6-phosphate overlay blotting with iodinated cation-independent mannose-6 phosphate receptor, Sleat and colleagues (1998) could not identify a protein specifically associated with the CLN4 disease process. However, two proteins were highly elevated (11.9-fold and 15.0-fold) compared with controls, which might suggest that their increased levels could represent a compensatory mechanism mobilized by the cell in response to the CLN4 disease process. Isolation of the gene associated with CLN4 and characterization of the structure and function of its product is a condition necessary to understand the pathogenic events underlying this rare form of NCL.

Acknowledgments

The authors wish to thank Dr. J. Tyynelä for kindly providing material for documentation, and Drs. S. Hofmann and E. Kominami for their gift of antibodies to PPT1 and subunit c of mitochondrial ATP synthase. The editorial assistance of Mrs. Maureen Stoddard Marlow is appreciated.

This work was supported in part by National Institutes of Health grant NS38988, the Batten's Disease Support and Research Association, and the New York State Office for Mental Retardation and Developmental Disabilities.

References

Banerjee, P., Dasgupta, A., Siakotos, A. N., and Dawson, G. (1992). Evidence for lipase abnormality: High levels of free and triacylglycerol forms of unsaturated fatty acids in neuronal ceroid-lipofuscinosis tissue. *Am. J. Med. Genet.* **42,** 549–554.

Beers, M. F. (1996). Inhibition of cellular processing of surfactant protein C by drugs affecting intracellular pH gradients. *J. Biol. Chem.* **71,** 14361–14370.

Berkovic, S. F., Carpenter, S., Andermann, F., Andermann, E., and Wolfe, L. S. (1988). Kufs' disease: A critical reappraisal. *Brain* **111,** 27–62.

Bizzozero, O. A. (1997). The mechanism and functional roles of protein palmitoylation in the nervous system. *Neuropediatrics* **28,** 23–26.

Braak, H., and Goebel, H. H. (1979). Pigmentoarchitectonic pathology of the isocortex in juvenile neuronal ceroid lipofuscinosis: Axonal enlargement in layer IIIab and cell loss in layer V. *Acta Neuropathol. (Berl.)* **46,** 79–83.

Braulke, T., Geuze, H. J., Slot, J. W., Hasilik, A., and von Figura, K. (1987). On the effects of weak bases and monensin on sorting and processing of lysosomal enzymes in human cells. *Eur. J. Cell. Biol.* **43,** 316–321.

Broom, M. F., Zhou, C., Broom, J. E., Barwell, K. J., Jolly, R. D., and Hill, D. F. (1998). The locus for ovine ceroid lipofuscinosis is syntenic to the human Batten disease variant CLN6. *J. Med. Genet.* **35,** 717–721.

Brück, W., and Goebel, H. H. (1998). Microglia activation in neuronal ceroid-lipofuscinosis. *Clin. Neuropathol.* **5,** 276.

Butor, C., Griffiths, G., Aronson, N. N., and Varki, A. (1995). Colocalization of hydrolytic enzymes with widely disparate pH optima: Implications for the regulation of lysosomal pH. *J. Cell Sci.* **108,** 2213–2219.

Buzy, A., Ryan, E. M., Jennings, K. R., Palmer, D. N., and Griffiths, D. E. (1996). Use of electrospray ionization mass spectrometry and tandem mass spectrometry to study binding of F0 inhibitors to ceroid lipofuscinosis protein, a model system for subunit c of mitochondrial ATP synthase. *Rapid Commun. Mass Spectrom.* **10,** 790–796.

Camp, L. A., and Hofmann, S. L. (1993). Purification and properties of a palmitoyl-protein thioesterase that cleaves palmitate from H-ras. *J. Biol. Chem.* **268,** 22566–22574.

Camp, L. A., Verkruyse, L. A., Afendis, S. J., Slaughter, C. A., and Hofmann, S. L. (1994). Molecular cloning and expression of palmitoyl-protein thioesterase. *J. Biol. Chem.* **269,** 23212–23219.

Chang, M. H., Bindloss, C. A., Grabowski, G. A., Qi, X., Winchester, B., Hopwood, J. J., and Meikle, P. J. (2000). Saposins A, B, C, and D in plasma of patients with lysosomal storage disorders. *Clin. Chem.* **46,** 167–174.

Christomanou, H., Chabas, A., Pampols, T., and Guardiola, A. (1989). Activator protein deficient Gaucher's disease. *Klin. Wochenschr.* **67,** 999–1003.

Croopnick, J. B., Choi, H. C., and Muller, D. M. (1998). The subcellular location of the yeast *Saccharomyces cerevisiae* homologue of the protein defective in juvenile form of Batten disease. *Biochem. Biophys. Res. Commun.* **250,** 335–341.

Das, A. S., Becerra, C. H. R., Yi, W., Lu, J-Y., Siakotos, A. N., Wisniewski, K. E., and Hofman, S. L. (1998). Molecular genetics of palmitoyl-protein thioesterase deficiency in the U.S. *J. Clin. Invest.* **102,** 361–370.

Das, A. M., Jolly, R. D., and Kohlschütter, A. (1999). Anomalies of mitochondrial ATP synthase regulation in four different types of neuronal ceroid lipofuscinosis. *Mol. Genet. Metab.* **66,** 349–355.

Dawson, G., Cho, S., Siakotos, A. N., and Kilkus, J. (1997). Low molecular weight storage material in infantile ceroid lipofuscinosis (CLN1). *Neuropediatrics* **28,** 31–32.

Doebber, T. W., Divor, A. R., and Ellis, S. (1978). Identification of a tripeptidyl aminopeptidase in the anterior pituitary gland: Effect on the chemical and biological properties of rat and bovine growth hormones. *Endocrinology* **103,** 1794–1804.

Dunphy, J. T., and Linder, M. E. (1998). Signalling functions of protein palmitoylation. *Biochim. Biophys. Acta.* **1436,** 245–261.

Dyer, M. R., and Walker, J. E. (1993). Sequences of members of the human gene family for the c subunit of mitochondrial ATP synthase. *Biochem. J.* **293,** 51–64.

Elleder, M., and Tyynelä, J. (1998). Incidence of neuronal perikaryal spheroids in neuronal ceroid lipofuscinoses (Batten disease). *Clin. Neuropathol.* **17,** 184–189.

Elleder, M., Drahota, Z., Lisa, V., Mares, V., Mandys, V., Muller, J., and Palmer, D. N. (1995). Tissue culture loading test with storage granules from animal models of neuronal ceroid-lipofuscinosis (Batten disease): Testing their lysosomal degradability by normal and Batten disease. *Am. J. Med. Genet.* **57,** 213–221.

Elleder, M., Sokolova, J., and Hrebicek, M. (1997). Follow-up study of subunit c of mitochondrial ATP synthase (SCMAS) in Batten disease and in unrelated lysosomal disorders. *Acta Neuropathol.* **93,** 379–390.

Ezaki, J., Wolfe, L. S., Higuti, T., Ishidoh, K., and Kominami, E. (1995). Specific delay of degradation of mitochondrial ATP synthase subunit c in late infantile neuronal ceroid lipofuscinosis (Batten disease). *J. Neurochem.* **64,** 733–741.

Ezaki, J., Wolfe, L. S., and Kominami, E. (1996). Specific delay in the degradation of mitochondrial ATP synthase subunit c in the late infantile neuronal ceroid lipofuscinosis is derived from cellular proteolysis dysfunction rather than structural alterations of subunit c. *J. Neurochem.* **67,** 1677–1687.

Ezaki, J., Wolfe, L. S., and Kominami, E. (1997). Decreased lysosomal subunit c-degrading activity in fibroblasts from patients with late infantile neuronal ceroid lipofuscinosis. *Neuropediatrics* **28,** 53–55.

Ezaki, J., Tanida, I., Kanehagi, N., and Kominami, E. (1999). A lysosomal proteinase, the late infantile neuronal ceroid lipofuscinosis gene (CLN2) product, is essential for degradation of a hydrophobic protein, the subunit c of ATP synthase. *J. Neurochem.* **72,** 2573–2582.

Ezaki, J., Takeda-Ezaki, M., Oda, K., and Kominami, E. (2000). Characterization of endopeptidase activity of tripeptidyl peptidase-I/CLN2 protein which is deficient in classical late infantile neuronal ceroid lipofuscinosis. *Biochem. Biophys. Res. Commun.* **268,** 904–908.

Faust, J., Rodman, J., Daniels, P., Dice, J., and Bronson, R. (1994). Two related proteolipids and dolichol-linked oligosaccharides accumulate in motor neuron degeneration mice (*mnd/mnd*), a model for neuronal ceroid-lipofuscinosis. *J. Biol. Chem.* **269,** 10150–10155.

Fearnley, I. M., Walker, J. E., Martinus, R. D., Jolly, R. D., Kirkland, K. B., Shaw, G. J., and Palmer, D. N. (1990). The sequence of the major protein stored in ovine ceroid lipofuscinosis is identical with that of the dicyclohexylcarbodiimide-reactive proteolipid of mitochondrial ATP synthase. *Biochem. J.* **268,** 751–758.

Fillingame, R. H. (1997). Coupling H^+ transport and ATP synthesis in F1F0-ATP synthases: Glimpses of interacting parts in a dynamic molecular machine. *J. Exp. Biol.* **200,** 217–224.

Forgac, M. (1998). Structure, function and regulation of the vacuolar (H+)-ATPases. *FEBS Lett.* **440,** 258–263.

Fukuda, M. (1994). Biogenesis of the lysosomal membrane. *Subcell. Biochem.* **22,** 199–230.

Goebel, H. H., and Braak, H. (1989). Adult neuronal ceroid-lipofuscinosis. *Clinical Neuropathol.* **8,** 109–119.

Goebel, H. H., Zeman, W., Patel, V. K., Pullarkat, R. K., and Lenard, H. (1979). On the ultrastructural diversity and essence of residual bodies in neuronal ceroid-lipofuscinosis. *Mech. Ageing Dev.* **10,** 73–70.

Golabek, A. A., Kaczmarski, W., Kida, E., Kaczmarski, A., Michalewski, M., and Wisniewski, K. E. (1999). Expression studies of CLN3 protein associated with a juvenile form of neuronal ceroid lipofuscinosis in mammalian cells. *Mol. Genet. Metab.* **66,** 277–282.

Golabek, A. A., Kida, E., Walus, M., Kaczmarski, W., Michalewski, M., and Wisniewski, K. E. (2000). CLN3 protein regulates lysosomal pH and alters intracellular processing of Alzheimer's amyloid-β protein precursor and cathepsin D in human cells. *Mol. Genet. Metab.* **70,** 203–213.

Haltia, M. J., Rapola, P., and Santavuori, P. J. (1973). Infantile type of so-called neuronal ceroid-lipofuscinosis. Histological and electron microscopic studies. *Acta Neuropathol. (Berl.)* **26,** 157–170.

Haltia, M., Tyynelä, J., Hirvasniemi, A., Herva, R., Ranta, U. S., and Lehesjoki, A.-E. (1999). CLN8. Northern epilepsy. *In* "The Neuronal Ceroid Lipofuscinoses (Batten Disease)" (H. H. Goebel *et al.*, eds.), pp. 117–124. IOS Press, Amsterdam, The Netherlands.

Hall, N. A., Lake, B. D., Dewji, N. N., and Patrick, A. D. (1991a). Lysosomal storage of subunit c of mitochondrial ATP synthase in Batten's disease (ceroid-lipofuscinosis). *Biochem. J.* **275,** 269–272.

Hall, N. A., Lake, B. D., and Patrick, A. D. (1991b). Recent biochemical and genetic advances in our understanding of Batten's disease (ceroid-lipofuscinosis). *Dev. Neurosci.* **13,** 339–344.

Hellsten, E., Vesa, J., Olkkonen, V. W., Jalanko, A., and Peltonen, L. (1996). Human palmitoyl protein thioesterase: Evidence for lysosomal targeting of the enzyme and disturbed cellular routing in infantile neuronal ceroid lipofuscinosis. *EMBO J.* **15,** 5240–5245.

Hiesberger, T., Huttler, S., Rohlmann, A., Schneider, W., Sandhoff, K., and Herz, J. (1998). Cellular uptake of saposin (SAP) precursor and lysosomal delivery by the low density lipoprotein receptor-related protein (LRP). *EMBO J.* **17,** 4617–4625.

Hiraiwa, M., Martin, B. M., Kishimoto, Y., Conner, G., Tsuji, S., and O'Brien, J. S. (1997). Lysosomal proteolysis of prosaposin, the precursor of saposins (sphingolipid activator proteins): Its mechanism and inhibition by ganglioside. *Arch. Biochem. Biophys.* **341,** 17–24.

Hofmann, S. L., Lee, L. A., Lu, J-Y., and Verkruyse, L. A. (1997). Palmitoyl-protein thioesterase and the molecular pathogenesis of the infantile neuronal ceroid lipofuscinosis. *Neuropediatrics* **28,** 27–30.

Holtschmidt, H., Sandhoff, K., Kwon, H. Y., Harzer, K., Nakano, T., and Suzuki, K. (1991). Sulfatide activator protein: Alternative splicing generates three messages RNAs and a newly found mutation responsible for a clinical disease. *J. Biol. Chem.* **266,** 7556–7560.

Hughes, S. M., Kay, G. W., Jordan, T. W., Rickards, G. K., and Palmer, D. N. (1999). Disease-specific pathology in neurons cultured from sheep affected with ceroid lipofuscinosis. *Mol. Genet. Metab.* **66,** 381–386.

Hunziker, W., and Geuze, H. J. (1996). Intracellular trafficking of lysosomal membrane protein. *BioEssays* **18,** 379–389.

Inui, K., and Wenger, D. A. (1983). Concentration of activator protein for sphingolipid hydrolysis in liver and brain samples from patients with lysosomal storage disorders. *J. Clin. Invest.* **72,** 1622–1628.

Inui, K., Kao, F. T., Fujibayashi, S., Jones, C., Morse, H. G., Law, M. L., and Wenger, D. A. (1985). The gene coding for sphingolipid activator protein, SAP-1, is on human chromosome 10. *Hum. Genet.* **69,** 197–200.

The International Batten Disease Consortium. (1995) Isolation of a novel gene underlying Batten disease, CLN3. *Cell* **82,** 949–957.

Isosomppi, J., Heinonen, O., Hiltunene, J. O., Greene, N. D. E., Vesa, J., Uusitalo, A., Mitchison, H. M., Saarma, M., Jalanko, A., and Peltonen, L. (1999). Developmental expression of palmitoyl protein thioesterase in normal mice. *Dev. Brain Res.* **118,** 1–11.

Janes, R. W., Munroe, P. B., Mitchison, H. M., Gardiner, R. M., Mole, S. E., and Wallace, B. A. (1996). A model for Batten disease protein CLN3: Functional implications from homology and mutations. *FEBS Lett.* **399,** 75–77.

Järvelä, I., Sainio, M., Rantamaki, T., Olkkonen, V. M., Carpen, O., Peltonen, L., and Jalanko, A. (1998). Biosynthesis and intracellular targeting of the CLN3 protein defective in Batten disease. *Hum. Mol. Genet.* **7,** 85–90.

Järvelä, I., Lehtovirta, M., Tikkanen, R., Kyttälä, A., and Jalanko, A. (1999). Defective intracellular transport of CLN3 is the molecular basis of Batten disease (JNCL). *Hum. Mol. Genet.* **8,** 1091–1098.

Junaid, M. A., Sklower Brooks, S., Wisniewski, K. E., and Pullarkat, R. K. (1999). A novel assay for lysosomal pepstatin-insensitive proteinase and its application for the diagnosis of late-infantile neuronal ceroid lipofuscinosis. *Clin. Chim. Acta* **281,** 169–176.

Junaid, M. A., Wu, G., and Pullarkat, R. K. (2000). Purification and characterization of bovine brain lysosomal pepstatin-insensitive proteinase, the gene product deficient in the human late-infantile neuronal ceroid lipofuscinosis. *J. Neurochem.* **74,** 287–294.

Kaczmarski, W., Wisniewski, K. E., Golabek, A. A., Kaczmarski, A., Kida, E., and Michalewski, M. (1999). Studies of membrane association of CLN3 protein. *Mol. Genet. Metab.* **66,** 261–264.

Katz, M. L., Gao, C-L., Prabhakaram, M., Shibuya, H., Liu, P-C., and Johnson, G. S. (1997). Immunochemical localization of the Batten disease (CLN3) protein in retina. *Invest. Ophthalmol. Vis. Sci.* **38,** 2375–2386.

Katz, M. L., Shibuya, H., Liu, P. C., Kaur, S., Gao, C. L., and Johnson, G. S. (1999). A mouse gene knockout model for juvenile ceroid-lipofuscinosis (Batten disease). *J. Neurosci. Res.* **57,** 551–556.

Kay, G. W., Hughes, S. M., and Palmer, D. N. (1999). In vitro culture of neurons from sheep with Batten disease. *Mol. Genet. Metab.* **67,** 83–88.

Kida, E., Wisniewski, K. E., and Golabek, A. A. (1993). Increased expression of subunit c of mitochondrial ATP synthase in brain tissue from neuronal ceroid lipofuscinoses and mucopolysaccharidosis cases but not in long-term fibroblast culture. *Neurosci. Lett.* **164,** 121–124.

Kida, E., Wisniewski, K. E., and Connell, F. (1995). Topographic variabilities of immunoreactivity to subunit c of mitochondrial ATP synthase and lectin binding in late infantile neuronal ceroid lipofuscinosis. *Am. J. Med. Genet.* **57,** 182–186.

Kida, E., Kaczmarski, W., Golabek, A. A., Kaczmarski, A., Michalewski, M., and Wisniewski, K. E. (1999). Analysis of intracellular distribution and trafficking of the CLN3 protein in fusion with the green fluorescent protein. *in vitro Mol. Genet. Metab.* **66,** 265–271.

Kieseier, B. C., Wisniewski, K. E., Park, E., Schuller-Levis, G., Mehta, P. D., and Goebel, H. H. (1997). Leukocytes in neuronal ceroid-lipofuscinosis: Function and apoptosis. *Brain & Develop.* **19,** 317–322.

Kitaguchi, T., Wisniewski, K. E., Maslinski, S., Maslinska, D., Wisniewski, T. M., and Kim, K. S. (1990). β-Protein immunoreactivity in brains of patients with neuronal ceroid lipofuscinosis: Ultrastructural and biochemical demonstration. *Neurosci Lett.* **112,** 155–160.

Klausner, R. D., and Sitia, R. (1990). Protein degradation in the endoplasmic reticulum. *Cell* **62,** 611–614.

Kominami, E., Ezaki, J., Muno, D., Ishido, K., Ueno, T., and Wolfe, L. S. (1992). Specific storage of subunit c of mitochondrial ATP synthase in lysosomes of neuronal ceroid lipofuscinosis (Batten's disease). *Biochem. J.* **111,** 278–282.

Kremmidiotis, G., Lensink, I. L., Bilton, R. L., Woollatt, E., Chataway, T. K., Sutherland, G. R., and Callen, D. F. (1999). The Batten disease gene product (CLN3p) is a Golgi integral membrane protein. *Hum. Mol. Genet.* **8,** 523–531.

Lake, B. D., and Rowan, S. A. (1997). Light and electron microscopic studies on subunit c in cultured fibroblasts in late infantile and juvenile Batten disease. *Neuropediatrics* **28,** 56–59.

Lane, S. C., Jolly, R. D., Schmechel, D. E., Alroy, J., and Boustany, R. M. (1996). Apoptosis as the mechanism of neurodegeneration in Batten's disease. *J. Neurochem.* **67,** 677–683.

Lauronen, L., Munroe, P. B., Järvelä, I., Autti, T., Mitchison, H. M., O'Rawe, A. M., Gardiner, R. M., Mole, S. E., Puranen, J., Hakkinen, A. M., and Santavuori, P. (1999). Delayed classic and protracted phenotypes of compound heterozygous juvenile neuronal ceroid lipofuscinosis. *Neurology* **52,** 360–365.

Leonova, T., Qi, X., Bencosme, A., Ponce, E., Sun, Y., and Grabowski, G. A. (1996). Proteolytic processing of prosaposin in insect and mammalian cells. *J. Biol. Chem.* **271,** 17312–17320.

Lu, J.-Y., Verkruyse, L. A., and Hofmann, S. L. (1996). Lipid thioesters derived from acylated proteins accumulate in infantile neuronal ceroid lipofuscinosis: Correction of the defect in lymphoblasts by recombinant palmitoyl-protein thioesterase. *Proc. Natl. Acad. Sci. (USA)* **93,** 10046–10050.

Marks, M. S., Woodruff, L., Ohno, H., and Bonifacino, J. S. (1996). Protein targeting by tyrosine- and di-leucine-based signals: Evidence for distinct saturable components. *J. Cell Biol.* **135,** 341–354.

Martin, J.-J., Libert, J., and Ceuterick, C. (1987). Ultrastructure of brain and retina in Kufs' disease (Adult type of ceroid-lipofuscinosis). *Clin. Neuropathol.* **6,** 381–394.

Martin, J.-J., Gottlob, I., Goebel, H. H., and Mole, S. E. (1999). CLN4. Adult NCL. *In* "The Neuronal Ceroid Lipofuscinoses (Batten Disease)" (H. H. Goebel *et al.*, eds.), pp. 77–90. IOS Press, Amsterdam, The Netherlands.

Mason, R. W. (1996). *In* "Subcellular Biochemistry" (Biology of the Liposome) (Lloyd and Mason, eds.), vol. **27,** pp. 159–190. Plenum Press, New York.

McGeoch, J. E., and Guidotti, G. (1997). A 0. 7-700 Hz current through a voltage-clamped pore: Candidate protein for initiator of neural oscillations. *Brain Res.* **766,** 188–194.

McGeoch, J. E. M., and Palmer, D. N. (1999). Ion pores made of mitochondrial ATP synthase subunit c in the neuronal plasma membrane and Batten disease. *Mol. Genet. Metab.* **66,** 387–392.

Michalewski, M., Kaczmarski, W., Golabek, A. A., Kida, E., Kaczmarski, A., and Wisniewski, K. E.

(1998). Evidence for phosphorylation of CLN3 protein associated with Batten disease. *Biochem. Biophys. Res. Commun.* **253,** 458–462.

Mitchison, H. M., Hofmann, S. L., Becerra, C. H. R., Munroe, P. B., Lake, B. D., Crow, Y. J., Stephenson, J. B., Williams, R. E., Hofman, I. L., Taschner, P. E. M., Martin, J. J., Philippart, M., Andermann, E., Andermann, F., Mole, S., Gardiner, R. M., and O'Rawe, A. M. (1998). Mutations in palmitoyl-protein thioesterase gene (PPT1; CLN1) causing juvenile neuronal ceroid lipofuscinosis with granular osmiophilic deposits. *Hum. Mol. Genet.* **7,** 291–297.

Mitchison, H. M., Bernard, D. J., Greene, N. D., Cooper, J. D., Junaid, M. A., Pullarkat, R. K., de Vos, N., Breuning, M. H., Owens, J. EW., Mobley, W. C., Gardiner, R. M., Lake, B. D., Taschner, P. E., and Nussbaum, R. L. (1999). Targeted disruption of the CLN3 gene provides a mouse model for Batten disease. *Neurobiol. Dis.* **6,** 321–334.

Medd, S. M., Walker, J. E., and Jolly, R. D. (1993). Characterization of the expressed genes for subunit c of mitochondrial ATP synthase in sheep with ceroid lipofuscinosis. *Biochem. J.* **293,** 65–73.

Mole, S. E., Mitchison, H. M., and Munroe, P. B. (1999). Molecular basis of the neuronal ceroid lipofuscinoses: Mutations in CLN1, CLN2, CLN3, and CLN5. *Hum. Mutat.* **14,** 199–215.

Munroe, P. B., Mitchison, H. M., O'Rawe, A. M., Anderson, J. W., Boustany, R-M., Lerner, T. J., Taschner, P. E. M., de Vos, N., Breuning, M. H., Gardiner, R. M., and Mole, S. E. (1997). Spectrum of mutations in the Batten disease gene, CLN3. *Am. J. Hum. Genet.* **6,** 1310–316.

Ng Ying Kin, N. M. K., Palo, J., Haltia, M., and Wolfe, L. S. (1983). High levels of brain dolichols in neuronal ceroid-lipofuscinosis and senescence. *J. Neurochem.* **40,** 1465–1473.

O'Brien, J. S., and Kishimoto, Y. (1991). Saposin proteins: Structure, function, and role in human lysosomal storage disorders. *FASEB J.* **5,** 301–308.

O'Brien, J. S., Carson, G. S., Seo, H. C., Hiraiwa, M., and Kishimoto, Y. (1994). Identification of prosaposin as a neurotrophic factor. *Proc. Natl. Acad. Sci. (USA)* **91,** 9593–9596.

Oka, A., Kurachi, Y., Mizuguchi, M., Hayashi, M., and Takashima, S. (1998). The expression of late infantile neuronal ceroid lipofuscinosis (CLN2) gene product in human brain. *Neurosci. Lett.* **257,** 113–115.

Page, A. E., Fuller, K., Chambers, T. J., and Warburton, M. J. (1993). Purification and characterization of a tripeptidyl peptidase from human osteoclastomas: evidence for its role in bone resorption. *Arch. Biochem. Biophys.* **306,** 354–359.

Palmer, D. N., Martinus, R. D., Cooper, S. M., Midwinter, G. G., Reid, J. C., and Jolly, R. D. (1989). Ovine ceroid lipofuscinosis. The major lipopigment protein and the lipid-binding subunit of mitochondrial ATP synthase have the same NH2-terminal sequence. *J. Biol. Chem.* **264,** 5736–5740.

Palmer, D. N., Fearnley, I. M., Walker, J. E., Hall, N. A., Lake, B. D., Wolfe, L. S., Haltia, M., Martinus, R. D., and Jolly, R. D. (1992). Mitochondrial ATP synthase subunit c storage in the ceroid-lipofuscinoses (Batten's disease). *Am. J. Med. Genet.* **42,** 561–567.

Palmer, D. N., Bayliss, S. L., Clifton, P. A., and Grant, V. J. (1993). Storage bodies in the ceroid-lipofuscinoses (Batten disease): Low-molecular-weight components, unusual amino acids and reconstitution of fluorescent bodies from non-fluorescent components. *J. Inherit. Metab. Dis.* **16,** 292–295.

Palmer, D. N., Jolly, R. D., van Mil, H. C., Tyynelä, J., and Westlake, V. J. (1997a). Different pattern of hydrophobic protein storage in different forms of neuronal ceroid lipofuscinosis (NCL, Batten disease). *Neuropediatrics* **28,** 45–48.

Palmer, D. N., Tyynelä, J., van Mil, H. C., Westlake, V. J., and Jolly, R. D. (1997b). Accumulation of sphingolipid activator proteins (SAPs) A and D in granular osmiophilic deposits in miniature Schnauzer dogs with ceroid-lipofuscinosis. *J. Inher. Metab. Dis.* **20,** 74–84.

Pardo, C. A., Rabin, B. A., Palmer, D. N., and Price, D. L. (1994). Accumulation of the adenosine triphosphate synthase subunit c in the mnd mutant mouse: A model for neuronal ceroid lipofuscinosis. *Am. J. Pathol.* **144,** 829–835.

Pearce, D. A., and Sherman, F. (1998). A yeast model for the study of Batten disease. *Proc. Natl Acad. Sci. (USA)* **95,** 6915–6918.

Pearce, D. A., Ferea, T., Nosel, S. A., Das, B., and Sherman, F. (1999a). Action of BTN1, the yeast orthologue of the gene mutated in Batten disease. *Nat. Gen.* **22,** 55–58.

Pearce, D. A., Carr, CJ, Das, B., and Sherman, F. (1999b). Phenotypic reversal of the btn1 defects in yeast by chloroquine: A yeast model for Batten disease. *Proc. Natl. Acad. Sci. (USA)* **96,** 11341–11345.

Poorthuis, B. J. H. M., Wevers, R. A., Kleijer, W. J., Groener, J. E. M., de Jong, J. G. N., van Weely, S., Niezen-Koning, K. E., and van Diggelen, O. P. (1999). The frequency of lysosomal storage diseases in The Netherlands. *Hum. Genet.* **105,** 151–156.

Puranam, K. L., Guo, W-X., Qian, W-H., Nikbakht, K., and Boustany, R. M. (1999). CLN3 defines a novel antiapoptotic pathway operative in neurodegeneration and mediated by ceramide. *Mol. Genet. Metab.* **66,** 294–308.

Qi, X., Kondoh, K., Krusling, D., Kelso, G. J., Leonova, T., and Grabowski, G. A. (1999). Conformational and amino acid residue requirements for the saposin C neuritogenic effect. *Biochemistry* **38,** 6284–6291.

Ranta, S., Zhang, Y., Ross, B., Lonka, L., Takkunen, E., Messer, A., Sharp, J., Wheeler, R., Kusumi, K., Mole, S., Liu, W., Soares, M. B., Bonaldo, M. F., Hirvasniemi, A., de la Chapelle, A., Gilliam, T. C., and Lehesjoki, A. E. (1999). The neuronal ceroid lipofuscinoses in human EPMR and *mnd* mutant mice are associated with mutations in CLN8. *Nature* **23,** 233–236.

Rawlings, N. D., and Barrett, A. J. (1999). Tripeptidyl-peptidase I is apparently the CLN2 protein absent in classical late-infantile neuronal ceroid lipofuscinosis. *Biochim. Biophys. Acta* **1429,** 496–500.

Reitman, M. L., and Kornfeld, S. (1981). UDP-N-acetylglucosamine:glycoprotein N-acetylglucosamine-1-phosphotransferase. Proposed enzyme for the phosphorylation of the high mannose oligosaccharide units of lysosomal enzymes. *J. Biol. Chem.* **256,** 4275–4281.

Rohrer, J., Schweizer, A., Russell, D., and Kornfeld, S. (1996). The targeting of lamp1 to lysosomes is dependent on the spacing of its cytoplasmic tail tyrosine sorting motif relative to the membrane. *J. Cell. Biol.* **132,** 565–576.

Sandhoff, K., Kolter, T., and van Echten-Deckert, G. (1995). Sphingolipid metabolism. Sphingolipid analogs, sphingolipid activator proteins, and the pathology of the cell. *Ann. N.Y. Acad. Sci.* **845,** 139–151.

Sandoval, I. V., Arredondo, J. J., Alcalde, J., Noriega, A. G., Vandekerckove, J., Jimenez, M. A., and Rico, M. (1994). The residues Leu(Ile)^{475}-Ile(Leu,Val,Ala)^{476}, contained in the extended carboxyl cytoplasmic tail, are critical for targeting of the resident lysosomal membrane protein LIMP II to lysosomes. *J. Biol. Chem.* **269,** 6622–6631.

Santavuori, P., Rapola, J., Haltia, M., Tyynelä, J., Peltonen, L., and Mole, S. E. (1999). CLN5. Finnish variant late infantile NCL. *In* "The Neuronal Ceroid Lipofuscinoses (Batten Disease)" (H. H. Goebel *et al.*, eds.), pp. 91–101. IOS Press, Amsterdam, The Netherlands.

Savukoski, M., Kestilä, M., Williams, R., Järvelä, I., Sharp, J., Harris, J., Santavuori, P., Gardiner, M., and Peltonen, L. (1994). Defined chromosomal assignment of CLN5 demonstrates that at least four genetic loci are involved in the pathogenesis of human ceroid lipofuscinosis. *Am. J. Hum. Genet.* **55,** 695–701.

Savukoski, M., Klockars, T., Holmberg, V., Santavuori, P., Lander, E. S., and Peltonen, L. (1998). CLN5, a novel gene encoding a putative transmembrane protein mutated in Finish variant late infantile neuronal ceroid lipofuscinosis. *Nat. Gen.* **19,** 286–288.

Sharp, J., Wheeler, R. B., Lake, B. D., Fox, M., Gardiner, R. M., and Williams, R. E. (1999). Genetic and physical mapping of the CLN6 gene on chromosome 15q21-23. *Mol. Genet. Metab.* **66,** 329–331.

Seitelberger, F., Jacob, H., and Schnabel, R. (1967). The myoclonic variant of cerebral lipidosis. *In*

"Inborn Disorders of Sphingolipid Metabolism" (S. M. Aronson and B. W. Volk, eds.), pp 43–74. Oxford, Pergamon Press.

Sleat, D. E., Sohar, I., Lackland, H., Majercak, J., and Lobel, P. (1996). Rat brain contains high levels of mannose-6-phosphorylated glycoproteins including lysosomal enzymes and palmitoyl-protein thioesterase, an enzyme implicated in infantile neuronal lipofuscinosis. *J. Biol. Chem.* **271,** 19191–19198.

Sleat, D. E., Donnelly, R. J., Lackland, H., Liu, C-G., Sohar, I., Pullarkat, R. K., and Lobel, P. (1997). Association of mutations in a lysosomal protein with classical late-infantile neuronal ceroid lipofuscinosis. *Science* **277,** 1802–1805.

Sleat, D. E., Sohar, I., Pullarkat, P. S., Lobel, P., and Pullarkat, R. K. (1998). Specific alteration in levels of mannose 6-phosphorylated glycoproteins in different neuronal ceroid lipofuscinoses. *Biochem. J.* **334,** 547–551.

Sleat, D. E., Gin, R. M., Sohar, I., Wisniewski, K. E., Sklower Brooks, S., Pullarkat, R. K., Palmer, D. N., Lerner, T. J., Boustany, R. M., Uldall, P., Siakotos, A. N., Donnelly, R. J., and Lobel, P. (1999). Mutational analysis of the defective protease in classic late-infantile neuronal ceroid lipofuscinosis, a neurodegenerative lysosomal storage disorder. *Am. J. Hum. Genet.* **64,** 1511–1523.

Sohar, I., Sleat, D. E., Jadot, M., and Lobel, P. (1999). Biochemical characterization of a lysosomal protease deficient in classical late infantile neuronal ceroid lipofuscinosis (LINCL) and development of an enzyme-based assay for diagnosis and exclusion of LINCL in human specimens and animal models. *J. Neurochem.* **73,** 700–711.

Suopanki, J., Tyynelä, J., Baumann, M., and Haltia, M. (1999). Palmitoyl-protein thioesterase, an enzyme implicated in neurodegeneration, is localized in neurons and is developmentally regulated in rat brain. *Neurosci. Lett.* **265,** 53–56.

Tahvanainen, E., Ranta, S., Hirvasniemi, A, Karila, E., Leisti, J., Sistonen, P., Weissenbach, J., Lehesjoki, A. E., and de la Chapelle, A. (1994). The gene for a recessively inherited human childhood progressive epilepsy with mental retardation maps to the distal short arm of chromosome 8. *Proc. Natl. Acad. Sci. (USA)* **91,** 7267–7270.

Tanner, A., and Dice, J. F. (1995). Batten disease fibroblasts in culture accumulate mitochondrial ATP synthase subunit 9. *Cell Biol. Int.* **19,** 71–75.

Tanner, A., Shen, B. H., and Dice, J. F. (1997). Turnover of F1F0-ATP synthase subunit 9 and other proteolipids in normal and Batten disease fibroblasts. *Biochim. Biophys. Acta* **1361,** 251–262.

Town, M., Jean, G., Cherqui, S., Attard, M., Forestier, L., Whitmore, S. A., Callen, D. F., Gribouval, O., Broyer, M., Bates, G. P., vant Hoff, W., and Antignac, C. (1998). A novel gene encoding an integral membrane protein is mutated in nephropathic cystinosis. *Nat. Genet.* **18,** 319–324.

Tsuboi, K., Hiraiwa, M., and O'Brien, J. S. (1998). Prosaposin prevents programmed cell death of rat cerebellar granule neurons in culture. *Brain Res. Dev. Brain Res.* **110,** 249–255.

Tyynelä, J., Palmer, D. N., Baumann, M., and Haltia, M. (1993). Storage of saposins A and D in infantile neuronal ceroid lipofuscinosis. *FEBS Lett.* **330,** 8–12.

Tyynelä, J., Baumann, M., Henseler, M., Sandhoff, K., and Haltia, M. (1995). Sphingolipid activator proteins in the neuronal ceroid-lipofuscinoses: An immunological study. *Acta Neuropathol. (Berl.)* **89,** 391–398.

Tyynelä, J., Suopanki, J., Baumann, M., and Haltia, M. (1997a). Sphingolipid activator proteins (SAPs) in neuronal ceroid lipofuscinoses (NCL). *Neuropediatrics* **28,** 49–52.

Tyynelä, J., Suopanki, J., Santavuori, P., Baumann, M., and Haltia, M. (1997b). Variant late infantile neuronal ceroid-lipofuscinosis: pathology and biochemistry. *J. Neuropathol. Exp. Neurol.* **56,** 369–375.

Vaccaro, A. M., Salvioli, R., Barca, A., Tatti, M., Ciaffoni, F., Maras, B., Siciliano, R., Zappacosta, F., Amoresano, A., and Pucci, P. (1995). Structural analysis of saposin C and B: Complete localization of disulfide bridges. *J. Biol. Chem.* **270,** 9953–9960.

2. Cellular Pathology and Pathogenic Aspects of NCLs 67

Vaccaro, A. M., Salvioli, R., Tatti, M., and Ciaffoni, F. (1999). Saposins and their interaction with lipids. *Neurochem. Res.* **24**, 307–314.

Verheijen, F. W., Verbeek, E., Aula, N., Beerens, C. E. M. T., Havelaar, A. C., Joosse, M., Peltonen, L., Aula, P., Galjaard, H., van der Spek, P. J., and Mancini, G. M. S. (2000). A new gene, encoding an anion transporter, is mutated in sialic acid storage diseases. *Nat. Genet.* **23**, 462–465.

Verkruyse, L. A., and Hofmann, S. L. (1996). Lysosomal targeting of palmitoyl-protein thioesterase. *J. Biol. Chem.* **26**, 15831–15836.

Vesa, J., Hellsten, E., Verkruyse, L. A., Camp, L. A., Rapola, J., Santavuori, P., Hofmann, S. L., and Peltonen, L. (1995). Mutations in the palmitoyl protein thioesterase gene causing infantile neuronal ceroid lipofuscinosis. *Nature* **376**, 584–587.

Vines, D., and Warburton, N. J. (1998). Purification and characterization of a tripeptidyl aminopeptidase I from rat spleen. *Biochim. Biophys. Acta* **1384**, 233–242.

Vines, D., and Warburton, N. J. (1999). Classical late infantile neuronal ceroid lipofuscinosis fibroblasts are deficient in lysosomal tripeptidyl peptidase I. *FEBS Lett.* **443**, 131–135.

von Heijne, G., and Gavel, Y. (1988). Topogenic signals in integral membrane proteins. *Eur. J. Biochem.* **174**, 671–678.

Waliany, S., Das, A. K., Gaben, A., Wisniewski, K. E., and Hofmann, S. L. (2000). Identification of three novel mutations of the palmitoyl-protein thioesterase (PPT) gene in children with neuronal ceroid-lipofuscinosis. *Hum. Mutat.* **2**, 206–207.

Williams, R. E., Lake, B. D., Elleder, M., and Sharp, J. D. (1999a). CLN6. Variant late infantile/early juvenile NCL. *In* "The Neuronal Ceroid Lipofuscinoses (Batten Disease)" (H. H. Goebel et al., eds.), pp. 102–114. IOS Press, Amsterdam, The Netherlands.

Williams, R. E., TopHu, M., Lake, B. D., Mitchell, W., and Mole, S. E. (1999b). CLN7. Turkish variant late infantile NCL. *In* "The Neuronal Ceroid Lipofuscinoses (Batten Disease)" (H. H. Goebel et al., eds.), pp. 114–115. IOS Press, Amsterdam, The Netherlands.

Wimer Mackin, S., and Granger, B. L. (1996). Transmembrane domain mutations influence the cellular distribution of lysosomal membrane glycoprotein A. *Biochem. Biophys. Res. Commun.* **229**, 472–478.

Wisniewski, K. E., Maslinska, D., Kitaguchi, T., Kim, K. S., Goebel, H. H., and Haltia, M. (1990a). Topographic heterogeneity of amyloid β-protein precursor epitopes in brains with various forms of neuronal ceroid lipofuscinoses suggesting defective processing of amyloid precursor protein. *Acta Neuropathol.* **80**, 26–30.

Wisniewski, K. E., Kida, E., Gordon Majszak, W., and Saitoh, T. (1990b). Altered amyloid β-protein precursor processing in brains of patients with neuronal ceroid lipofuscinosis. *Neurosci Lett.* **120**, 94–96.

Wisniewski, K. E., Kida, E., Patxot, O. F., and Connell, F. (1992). Variability in the clinical and pathological findings in the neuronal ceroid lipofuscinosis: Review of data and observations. *Am. J. Med. Genet.* **42**, 525–532.

Wisniewski, K. E., Kida, E., Connell, F., Elleder, M., Eviatar, L., and Konkol, R. J. (1993). New subform of the late infantile form of neuronal ceroid lipofuscinosis. *Neuropediatrics* **24**, 155–163.

Wisniewski, K. E., Golabek, A. A., and Kida, E. (1994). Increased urine concentration of subunit c of mitochondrial ATP synthase in neuronal ceroid lipofuscinoses patients. *J. Inherit. Metab. Dis.* **17**, 205–210.

Wisniewski, K. E., Kaczmarski, W., Golabek, A. A., and Kida, E. (1995). Rapid detection of subunit c of mitochondrial ATP synthase in urine as a diagnostic screening test for neuronal ceroid-lipofuscinosis. *Am. J. Med. Genet.* **57**, 246–249.

Wisniewski, K. E., Zhong, N., Kaczmarski, W., Kaczmarski, A., Sklower-Brooks, S., and Brown, W. T. (1998a). Studies of atypical JNCL suggest overlapping with other NCL forms. *Pediatr. Neurol.* **18**, 36–40.

Wisniewski, K. E., Zhong, N., Kaczmarski, W., Kaczmarski, A., Kida, E., Schwarz, K. O., Lazzarini,

A. M., Rubin, A. J., Stenroos, E. S., Johnson, W. G., and Wisniewski, T. M. (1998b). Compound heterozygous genotype is associated with protracted juvenile neuronal ceroid lipofuscinosis. *Ann. Neurol.* **42,** 106–110.

Wolfe, L. S., Kin, N. M., Baker, R. R., Carpenter, S., and Andermann, F. (1977). Identification of retinoyl complexes as the autofluorescent component of the neuronal storage material in Batten disease. *Science* **195,** 1360–1362.

Yan, W. L., Lerner, T. J., Haines, J. L., and Gusella, J. F. (1994). Sequence analysis and mapping of a novel human mitochondrial ATP synthase subunit 9 cDNA (ATP5G3). *Genomics* **24,** 375–377.

Zeman, W., and Dyken, P. (1969). Neuronal-ceroid lipofuscinosis (Batten's disesae): Relationship to amaurotic. idiocy? *Pediatrics* **44,** 570–583.

Zhong, N., Wisniewski, K. E., Kaczmarski, A., Ju, W., Xu, W., McLendon, L., Liu, B., Kaczmarski, W., Sklower Brooks, S., and Brown, W. T. (1998). Molecular screening of Batten disease: Identification of a missense mutation (E295K) in the CLN3 gene. *Hum. Genet.* **102,** 57–62.

3

Positional Candidate Gene Cloning of CLN1

Sandra L. Hofmann,* Amit K. Das, Jui-Yun Lu, and Abigail A. Soyombo
Department of Internal Medicine and the Hamon Center
for Therapeutic Oncology Research
University of Texas Southwestern Medical Center
Dallas, Texas 75390

I. Introduction
II. Linkage Disequilibrium Mapping of the CLN1 Locus in the Finnish Population
III. Palmitoyl-Protein Thioesterase Defines a New Pathway in Lysosomal Catabolism
IV. Enzymology of PPT
V. Posttranslational Processing and Lysosomal Targeting of PPT
VI. The Physiological Role of PPT
VII. Palmitoyl-Protein Thioesterase-2 (PPT2)
VIII. The PPT cDNA and Gene
IX. The Molecular Genetics of CLN1/PPT Deficiency
    A. PPT Deficiency as a Simple Recessive Genetic Trait
    B. Specific Mutations in PPT: Some Examples
X. Laboratory Diagnosis of PPT Deficiency
XI. Prospects for Cause-Specific Treatment of PPT Deficiency
    References

*Address for correspondence: E-mail: hofmann@simmons.swmed.edu

ABSTRACT

Mutations in the CLN1 gene encoding palmitoyl-protein thioesterase (PPT) underlie the recessive neurodegenerative disorder, infantile Batten disease, or infantile neuronal ceroid lipofuscinosis (INCL). The CLN1 gene was mapped to chromosome 1p32 in the vicinity of a microsatellite marker HY-TM1 in a cohort of Finnish INCL families, and mapping of the PPT gene to the CLN1 critical region (and the discovery of mutations in PPT in several unrelated families) led to conclusive identification of PPT as the disease gene. PPT is a lysosomal thioesterase that removes fatty acids from fatty-acylated cysteine residues in proteins. The accumulation of fatty acyl cysteine thioesters can be reversed in INCL cells by the exogenous administration of recombinant PPT, which enters the cells through the mannose 6-phosphate receptor pathway. Over two dozen PPT mutations have been found in PPT-deficient patients worldwide. In the United States, all PPT-deficient patients show "GROD" histology but the age of onset of symptoms is later in some children due to the presence of missense mutations that result in enzymes with residual PPT activity. Now that INCL is known to be caused by a defect in a soluble lysosomal enzyme, appropriate therapies may be forthcoming. Prospects for therapy include enzyme replacement, stem cell transplantation, gene therapy, and metabolic therapy aimed at depleting the abnormal substrate accumulation in the disease.

I. INTRODUCTION

Infantile Batten disease, or infantile neuronal ceroid lipofuscinosis (INCL) is caused by autosomal recessive mutations in a single gene, designated *CLN1*. The *CLN1* gene encodes a soluble lysosomal enzyme, palmitoyl-protein thioesterase (PPT), which hydrolyzes fatty acids from lipid-modified proteins. Prior to the mid-1990s, INCL was considered quite rare outside of Finland, where the disorder was first described, and physicians in the United States were reluctant to make the diagnosis unless the patient reported Finnish ancestry. Following the discovery of the *CLN1* gene in 1995, it is now apparent that mutations in *CLN1* account for between 20% and 25% of all cases of neuronal ceroid lipofuscinosis in the United States. Of U.S. patients with defects in *CLN1*, about half closely resemble the classical "Finnish" patients and, unfortunately, show the same fulminant course, with absence of higher cortical function by age 3. Somewhat surprisingly, the remaining half of U.S. patients with defects in *CLN1* follow a clinical course virtually indistinguishable from late-infantile or juvenile Batten patients. However, all *CLN1* patients, regardless of age of onset and pace of progression, display a common characteristic ultrastructural pathology, termed granular osmiophilic deposits, or GROD.

The neurodegenerative disorder caused by PPT deficiency is unique among many other lysosomal storage diseases in that neuronal loss, rather than neuronal storage, dominates the clinical picture. The major findings at autopsy are limited to the central nervous system, and in the final stages of the disease the brain is very small and atrophic, often weighing less than 500 g, a finding distinctly different from other lysosomal storage diseases that affect the brain. The cause of the severe neuronal loss, which is most prominent in neocortical areas, is one of the central mysteries surrounding INCL.

The discovery of the basic defect in INCL is a classic success story of the melding of the new "reverse" genetics (or positional cloning) with classical biochemistry to gain very rapid insight into a previously enigmatic biological process.

II. LINKAGE DISEQUILIBRIUM MAPPING OF THE *CLN1* LOCUS IN THE FINNISH POPULATION

The search for the *CLN1* gene began in 1989, with the collection of blood samples from members of 15 Finnish INCL families. A random scan of markers located at intervals across the human genome showed that *CLN1* was linked to two markers, D1S57 and D1S7, on chromosome 1p32 (Jarvela et al., 1991). The *CLN1* region was further refined by linkage analysis of an additional 35 INCL families, placing the gene within a 25-cM chromosomal region in the vicinity of the L-*myc* oncogene (Jarvela, 1991). A genetic technique well suited to the study of homogenous populations, termed linkage disequilibrium mapping, was applied to Finnish samples, and assigned the *CLN1* locus to a 300-kb region centromeric to L-*myc* (Hellsten et al., 1995). A physical map consisting of genomic DNA clones was constructed over this region and a number of novel transcripts in the *CLN1* region were identified. While the work to identify the *CLN1* gene was proceeding, the gene encoding a new hydrolytic enzyme (PPT) was mapped to chromosome region 1p32 by fluorescence in situ hybridization (FISH) to metaphase chromosomes (Schriner et al., 1996). This general chromosomal region agreed well with the previously reported location of *CLN1* and was therefore considered as a possible candidate for *CLN1*. Fine mapping of the PPT gene on the constructed physical map (Figure 3.1) revealed that several of the genomic clones in the critical *CLN1* region contained the *PPT* gene, and a high-resolution technique (Fiber-FISH) was then applied to locate the *PPT* gene precisely on the map by hybridizing the *PPT*-containing clones isolated from phage libraries (Vesa et al., 1995). *PPT* was found to be located approximately 200 kb centromeric from the L-*myc* gene. The finding of mutations in this gene in multiple INCL patients verified that the PPT gene and the CLN1 gene were one and the same (Vesa et al., 1995).

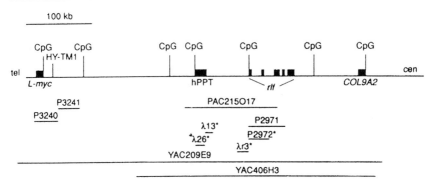

Figure 3.1. Physical map of the *CLN1* region at 1p32. Known genes in the region are indicated by boxes and genomic clones by horizontal bars. The human PPT gene was found to be located approximately 200 kb centromeric from the L-*myc* gene in the vicinity of the microsatellite marker HY-TM1, which had been shown to be in strong linkage disequilibrium with *CLN1*. [Reproduced from Vesa *et al.* (1995), with permission.]

III. PALMITOYL-PROTEIN THIOESTERASE DEFINES A NEW PATHWAY IN LYSOSOMAL CATABOLISM

Palmitoyl-protein thioesterase (PPT) was discovered during efforts to understand the mechanism of the turnover of fatty acids bound to proteins (Camp and Hofmann, 1993; Camp *et al.*, 1994). Many proteins associated with the plasma membrane, including those that transmit signals (such as hormonal signals) from outside the cell to the cell interior, are modified by fatty acids that are covalently attached to cysteine residues in the protein backbone (Figure 3.2). These fatty acids are crucial for membrane association and participate in protein–protein interactions that are important in signal transduction (Mumby, 1997). PPT was identified and characterized based on its ability to hydrolyze the 16-carbon fatty acid palmitate from the oncogene H-Ras in in vitro assays. Using the removal of palmitate from H-Ras as an assay for the enzyme, PPT was purified and the primary amino acid sequence was deduced from the cloned PPT cDNA (Camp and Hofmann, 1993; Camp *et al.*, 1994). From a comparison of the amino-terminal sequence of the mature protein with the amino acid sequence deduced from the cDNA, it was concluded that PPT was synthesized with a cleavable signal peptide. As signal peptides mediate the co-translational translocation of proteins across membranes in the endoplasmic reticulum, it seemed likely that PPT was either a secreted protein, or one that resides in an extracytoplasmic compartment within the cell, such as endoplasmic reticulum, Golgi, or lysosomes. The chromosomal localization of PPT1 to 1p32 (Schriner *et al.*, 1996), and its implication in a putative lysosomal storage disease (Vesa *et al.*, 1995), led to the appreciation of PPT's true role in the lysosomal degradation of S-acylated proteins.

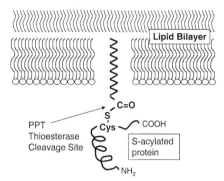

Figure 3.2. The thioesterase cleavage site of PPT. Palmitoylated proteins are normally found associated with the inner surface of the plasma membrane as shown. Fatty acids such as palmitate are often found in thioester linkage to cysteine residues in proteins. PPT removes fatty acids from proteins by hydrolyzing a thioester bond between the fatty acid and the cysteine residue. The natural substrates of PPT are fatty acylated proteins and peptides undergoing degradation in the lysosome.

IV. ENZYMOLOGY OF PPT

A number of studies have been performed on the purified PPT protein. The key points of these studies are summarized in Table 3.1. PPT readily hydrolyzes fatty acids from modified cysteine residues in several proteins, including H-Ras, G proteins, and palmitoylated albumin (Camp et al., 1994; Soyombo and Hofmann, 1997). Palmitoyl-CoA and synthetic compounds consisting of palmitate bound in thioester linkage to sugars are also substrates, indicating a lack of specificity for the "leaving" group in the thioesterase reaction (Camp et al., 1994; van Diggelen et al., 1999; Voznyi et al., 1999). Oxyesters do not appear to be substrates. Fatty acids from between 14 and 18 carbons are hydrolyzed at nearly equivalent rates, with a rather steep dropoff in activity outside this range (Camp et al., 1994). In reactions utilizing most protein substrates, the addition of detergent to the reaction mixtures is required, presumably to solubilize the substrate (Camp and Hofmann, 1993). Detergent is not required for the hydrolysis of the water-soluble substrate palmitoyl CoA (Camp et al., 1994). The pH optimum depends on the substrate used in the assay. PPT is more active against protein substrates at neutral pH, but some small molecule substrates are hydrolyzed at a pH optimum closer to 5.0. PPT is slowly inactivated by 1 mM DTT, probably because it contains several intramolecular disulfide bonds. It is inactivated by heavy metals and stabilized in the presence of EDTA, presumably due to the presence of neighboring sulfhydryls in disulfide linkage at amino acid residues 45 and 46 (Bellizzi et al., 1999). The substrate analog hexadecylsulfonyl fluoride irreversibly inactivates the enzyme and modifies the active site serine (Das et al., 2000). In contrast, the

Table 3.1. Properties of Palmitoyl-Protein Thioesterase

1. 37-kDa lysosomal thioesterase
2. Specific for long-chain fatty acyl thioesters with preference for 14–18 carbons
3. Substrates in vivo: fatty acid thioesters of cysteine-containing peptides.
4. Substrates in vitro: palmitoyl CoA, S-methylumbelliferyl-6S-palmitoyl-β-glucoside, S-fatty acylated peptides and proteins. Specific activity of recombinant PPT is ~1 μmol of palmitoyl-CoA hydrolyzed per minute per milligram of protein at pH 5. (Turnover number: 0.62/s).
5. Acidic to neutral pH optimum, depending on substrate; enzyme is unstable at alkaline pH
6. Inhibited by diethylpyrocarbonate and insensitive to phenylmethanesulfonyl fluoride and diisopropylfluorophosphate
7. Appears as a 37-kDa doublet (or triplet) by SDS-PAGE due to heterogenous glycoslyation. Has three asparagine-linked glycosylation sites. Glycosylation required for activity/stability.
8. Minimally processed—removal of signal peptide and N-terminal dipeptide. Fully active after signal peptide removal.
9. Catalysis via Ser-Asp-His triad. Structurally similar to bacterial lipases—α/β hydrolase fold.
10. Two disulfide bonds. Loses activity upon reduction. Sensitive to heavy-metal contamination of buffers. Stabilized by storage in EDTA.
11. Modified by mannose 6-phosphate; targeted to lysosomes via the mannose 6-P receptor
12. Ubiquitously expressed. Highest protein/enzyme activity in brain and testes.

enzyme is insensitive to the class-specific serine-modifying reagent PMSF due to geometric constraints surrounding the active site serine. Diethylpyrocarbonate rapidly inactivates the enzyme, indicating the presence of a crucial histidine residue (Camp and Hofmann, 1993).

PPT has been crystallized and its three-dimensional structure determined to 2.0Å resolution (Bellizzi et al., 1999) (Figure 3.3). The solved structure reveals an α/β fold, consisting of a central six-stranded parallel β sheet alternating with α helices. This structure is characteristic of a number of lipases and esterases (Schrag and Cygler, 1997). The closest structural neighbors to PPT are carboxylesterase from *Pseudomonas fluorescens* and lipase from *Candida antarctica*. Structurally, PPT most closely resembles the dienelactone hydrolase and haloperoxidase subfamilies because, like these two subfamilies, PPT has an insertion between β4 and αB that forms one domain of the protein. A second domain created by a large insertion between β6 and β7 (residues 140–223) forms most of the fatty acid-binding site (Bellizzi et al., 1999).

The catalytic mechanism of PPT can be inferred from its structure and is consistent with biochemical data. PPT utilizes a triad composed of serine, histidine, and aspartic acid. This "catalytic triad" of amino acids is utilized by many classic hydrolytic enzymes, such as chymotrypsin (Gold and Fahrney, 1964). In PPT, the members of the triad are Ser115, His289, and Asp233. Ser115, the catalytic nucleophile, is modified by palmitate in a structure that was solved using crystals grown in the presence of the substrate palmitoyl CoA (Bellizzi et al., 1999).

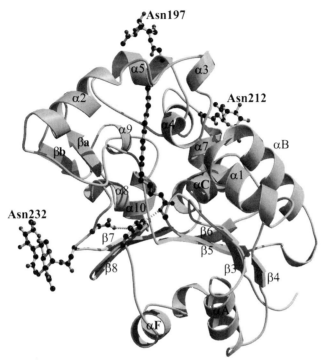

Figure 3.3. Three dimensional structure of PPT. PPT is a globular enzyme consisting of six parallel β strands alternating with α helices organized in a structure known as the α/β hydrolase fold typical of lipases. There is a large insertion between β6 and β7 (residues 140–223), which forms a second domain that forms most of the fatty acid-binding site. Catalytic active site residues Ser115, Asp233, and His289 are shown. [Reprinted from Bellizzi *et al.*, (1999) with permission. Copyright 1999 National Academy of Sciences U.S.A.]

V. POSTTRANSLATIONAL PROCESSING AND LYSOSOMAL TARGETING OF PPT

The posttranslational processing and lysosomal targeting of PPT1 has been well established (Hellsten *et al.*, 1996; Sleat *et al.*, 1996; Verkruyse and Hofmann, 1996; Verkruyse *et al.*, 1997). As a classical soluble lysosomal enzyme, PPT is synthesized on membrane-bound ribosomes and targeted to the lysosome through the mannose 6-phosphate receptor pathway, a hallmark of lysosomal enzyme trafficking (Hellsten *et al.*, 1996; Verkruyse and Hofmann, 1996). This scheme, reviewed in (Kornfeld, 1990), is presented in Figure 3.4.

PPT is synthesized on membrane bound ribosomes and is translocated into the lumen of the endoplamic reticulum in a process mediated by a 25 amino

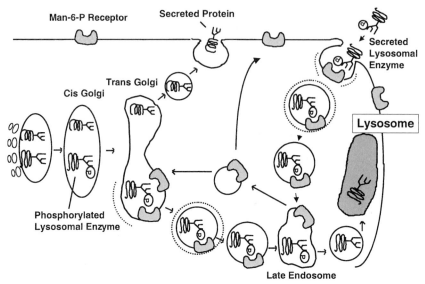

Figure 3.4. Lysosomal enzyme trafficking. PPT is synthesized on membrane-bound ribosomes and is phosphorylated (P) on oligosaccharide mannose residues in the *cis*-Golgi. The mannose 6-phosphate binds to a specific receptor (the mannose 6-phosphate receptor) located in the *trans*-Golgi. The mannose 6-phosphate receptor targets the newly synthesized PPT to the lysosome. Mannose 6-phosphate receptors are also found on the cell surface, where they can recapture mannose 6-phosphate-modified proteins and carry them to the lysosome. The mannose 6-phosphate receptor pathway provides a means to correct lysosomal storage disorders by enzyme replacement therapy. (Adapted from *Molecular Cell Biology* by Lodish *et al.* © 1986, 1990, 1996 by Scientific American Books, Inc. used with permission by W. H. Freeman and Company.)

acid signal peptide that is removed co-translationally (Camp *et al.*, 1994). In the Golgi network, three asparagine-linked glycosylation sites (at residues 197, 212, and 232) are modified by complex oligosaccharides (Bellizzi *et al.*, 1999). Like other proteins in this pathway, the earliest oligosaccharides are of the high-mannose type, which are sensitive to cleavage by endoglycosidase H (Verkruyse and Hofmann, 1996). Further processing of the oligosaccharides can be inferred from the observation that the mature protein purified from bovine brain contains some hybrid structures partially resistant to endoglycosidase H (Camp *et al.*, 1994).

Oligonucleotide-directed mutagenesis of the N-linked glycosylation sites in PPT has revealed that glycosylation is necessary for enzyme activity and/or stability. Mutant enzymes that are glycosylated at any two of the three sites (or at Asn232 alone) retain 75% or more of normal activity, while a mutant enzyme that is glycosylated at Asn212 alone retains about 50% of normal activity, and an enzyme in which all three glycosylation sites have been altered has no detectable activity (Bellizzi et al., 1999). Furthermore, when recombinant PPT is deglycosylated *in vitro* using PNGase F, it seems to precipitate from solution in largely inactive form (Verkruyse and Hofmann, unpublished results). Therefore, the oligosaccharides that modify PPT play an important role in maintaining PPT structure and function.

Oligosaccharides also play a special role in the "life cycle" of lysosomal enzymes, because these sugar residues are important for the next step in PPT maturation, in which PPT is recognized by a phosphotransferase that modifies one or more of the mannoses by phosphate (Hellsten et al., 1996; Verkruyse and Hofmann, 1996). It is this mannose 6-phosphate that serves as a signal for targeting PPT (and most lysosomal enzymes) to the lysosome (Kornfeld, 1990). Experimental confirmation of this process comes from the observation that fibroblasts from patients with the rare disorder I-cell disease (who lack the phosphotransferase) secrete all of their PPT into the tissue culture medium (Verkruyse et al., 1997). Interestingly, mannose 6-phosphate receptors are also found on the surface of cells, and like many soluble lysosomal enzymes, mannose 6-phosphate-modified PPT added to the cell exterior is taken up into the cell by receptor-mediated endocytosis after binding to the mannose 6-phosphate receptor (Figure 3.4). The mannose 6-phosphate receptor pathway therefore provides a means of delivery of exogenously added enzyme to the cell interior and serves as the basis for lysosomal enzyme replacement therapy. In fact, mannose 6-phosphate modified PPT has been added to INCL cells and shown to correct the metabolic defect in these cells (Lu et al., 1996).

In contrast to other lysosomal enzymes, activation by proteolytic processing does not seem to be a feature of the enzyme, as PPT is fully active following removal of the signal peptide (Camp et al., 1994). The mature enzyme appears as a doublet at 37 kDa/35 kDa on sodium dodecyl sulfate-polyacrylamide gel electrophoresis, due to heterogeneous glycosylation (Camp et al., 1994; Hellsten et al., 1996; Verkruyse and Hofmann, 1996). An additional two amino-terminal amino acids are present in the recombinant enzyme (His-Leu) that are not present in the mature enzyme purified from brain as determined by N-terminal amino acid sequencing. This minor processing, which is consistent with the action of a lysosomal dipeptidyl aminopeptidase on PPT, seems to be of little significance, as both unprocessed and amino-terminally processed forms are fully active (Camp et al., 1994).

VI. THE PHYSIOLOGICAL ROLE OF PPT

PPT was first purified by virtue of its ability to remove fatty acids from cysteine residues in proteins in vitro. A role for PPT in removing fatty acids from fatty acylated proteins in vivo has been demonstrated by metabolic labeling studies

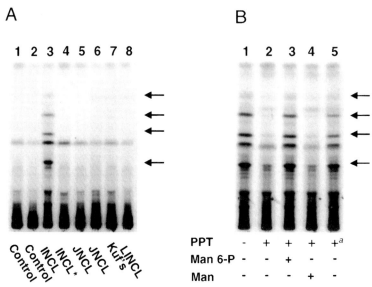

Figure 3.5. Identification of metabolites that accumulate in PPT-deficient cells. (A) INCL and control lymphoblasts were labeled with [^{35}S]cysteine, extracted with organic solvents, and subjected to thin-layer chromatography. At least four discrete compounds (arrows) are shown to be specific for INCL cells. The compounds have been shown to be cysteine thioesters of long-chain fatty acids (primarily palmitate), and presumably represent small fatty acylated peptides. (B) Correction of the metabolic defect in INCL cells by addition of mannose 6-phosphate-modified PPT to the cell culture medium. The correction is mediated through the mannose 6-phosphate receptor pathway because it is blocked in the presence of mannose 6-phosphate. [a]Recombinant PPT that is produced in the Sf9 expression system (and does not contain mannose 6-phosphate) is ineffective in correcting the metabolic abnormality. [Reprinted from Lu et al. (1996), with permission. Copyright 1996 National Academy of Sciences U.S.A.]

(Lu et al., 1996; Hofmann and Lu, 1999; Hofmann and Verkruyse, 1999). If cells from INCL patients are labeled with [^{35}S]cysteine, the label is incorporated into newly synthesized proteins. If lipid extracts are subsequently made from the cells after labeling, one sees specific accumulation of [^{35}S]cysteine-labeled compounds by thin-layer chromatography (Figure 3.5A). These compounds are specific to PPT-deficient patients and are not seen in normal controls or in cells derived from patients with other forms of Batten disease.

There is further evidence that the palmitate- and cysteine-labeled compounds are derived from acylated proteins undergoing degradation in the lysosome (Lu and Hofmann, manuscript in preparation). Their formation is blocked by cycloheximide, which blocks the incorporation of cysteine into protein, and by inhibitors of lysosomal function, such as chloroquine and leupeptin. In addition, the accumulating compounds largely co-migrate with the lysosomal fraction in density-gradient fractionation of subcellular organelles.

It is likely that the lipid thioesters that accumulate in INCL cells are toxic to neurons and that this toxicity is central to the mechanism of neurodegeneration in INCL. The accumulation of lipid thioesters in INCL cells can be prevented by the addition of mannose 6-phosphate-modified PPT to the tissue culture medium (Figure 3.5B). As discussed in the preceding section, the demonstration of correction of the metabolic defect in INCL cells by recombinant PPT serves as a rationale for enzyme replacement as a therapy for INCL.

VII. PALMITOYL-PROTEIN THIOESTERASE-2 (PPT2)

A second lysosomal thioesterase has been identified based on a 20% amino acid identity with PPT (Soyombo and Hofmann, 1997). Like PPT, it is ubiquitously expressed, but it has a distinct distribution within tissues. PPT2 is easily identified in human urine (Sleat *et al.*, 1997). Small long-chain fatty acid thioesters are excellent substrates for PPT2, but palmitoyl-proteins (at least those that are substrates for PPT) are not hydrolyzed. Careful studies suggest that PPT2 does not complement the metabolic defect of PPT-deficient cells (Soyombo and Hofmann, 1997). Therefore, the physiological function of PPT2 is unknown. The gene for PPT2 is located within the major histocompatibility complex III on chromosome 6p21.3. This location does not correspond to any other forms of NCL where the chromosomal location of the disease gene is known.

VIII. THE PPT cDNA AND GENE

The *PPT* gene spans approximately 25 kb and is comprised of nine exons (Figure 3.6A). cDNAs encoding human PPT have been characterized (Schriner *et al.*, 1996). A major 2.5-kb mRNA is detected in most tissues. The human PPT cDNA consists of a short 5′-untranslated region of 14 base pairs, an open reading frame of 918 base pairs, and a 1388-bp 3′ untranslated region. The amino acid sequence deduced from the cDNA contains 306 amino acids, of which the first 25 amino acids constitute a leader peptide (Figure 3.6B). Sequence motifs characteristic of thioesterases (Naggert *et al.*, 1988) (glycine-X-serine-X-glycine, and glycine-aspartic acid-histidine) are present at amino acid residues 113–117 and 287–289, respectively. Thioesterases contain the classical "catalytic triad" residues of serine, aspartic (or glutamic) acid, and histidine, as do many lipases and proteases (Lawson *et al.*, 1994; Witkowski *et al.*, 1994). Three potential asparagine-linked glycosylation sites are found near the carboxyl terminus of the protein at positions

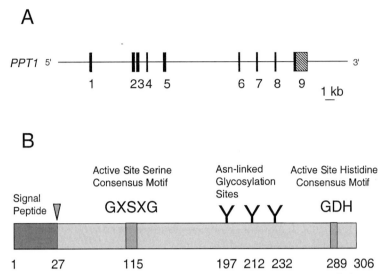

Figure 3.6. PPT gene and protein. (A) The PPT gene spans 25 kb and is comprised of nine exons. Exon 1 contains a very short 5′-untranslated region and exon 9 contains 1388 bp of 3′-untranslated region. A single 2.4-kb message is produced. (B) Schematic representation of PPT protein as deduced from the cDNA sequence. PPT is a single chain enzyme with a cleaved 25-amino acid signal peptide. The mature protein begins at amino acid residue 28. (A dipeptide is removed from the N-terminus without an effect on enzyme activity.) The locations of consensus motifs containing active site serine and histidine residues and three asparagine-linked glycosylation sites are shown.

197, 212, and 232. All of these sites are utilized to some degree in vivo (Bellizzi et al., 1999).

PPT mRNA is found ubiquitously in a variety of tissues from the human (Schriner et al., 1996) and rat (Camp et al., 1994), with fairly high and uniform levels in lung, spleen, brain, and testis, and relatively little mRNA in liver and skeletal muscle. The correlation between mRNA levels and levels of enzyme activity is imperfect, as much higher activity (and protein) levels are found in testis and brain as compared to other tissues (Camp and Hofmann, 1993). Interestingly, brain and testis are the two most severely affected organs histopathologically in INCL (Haltia et al., 1973).

Database searches reveal the presence of close PPT orthologs in mammals as well as in diverse organisms, such as *Drosophila* (fruit fly), *Caenorhabditis elegans* (roundworm), fission yeast, and even plants (*Arabidopsis* and others) (Das et al., 1998). These observations highlight the universal occurrence of protein palmitoylation in eukaryotic cells (which has been demonstrated in animals and plants) and presumably reflects the need for disposal and recycling of the digestion products resulting from this process.

IX. THE MOLECULAR GENETICS OF CLN1/PPT DEFICIENCY

A. PPT deficiency as a simple recessive genetic trait

Because PPT is a single-chain soluble lysosomal enzyme, the genetics of enzyme deficiency are relatively straightforward and easily understood as a simple recessive genetic trait. That is, the enzyme is encoded by two alleles (a maternal and a paternal) in each cell. Both must be defective to cause the disease. Carriers for PPT deficiency (for example, the parents of affected children) have one-half of the normal PPT activity. As is the case for most enzymatic deficiencies, 50% of normal activity is more than sufficient for normal cellular function.

There are a number of ways that the PPT gene can be defective and lead to PPT deficiency. Nonsense alleles (mutations that cause premature termination and truncation of the enzyme) contribute nothing to the overall enzymatic activity. Other mutations that create functionless or missing enzyme include frameshift mutations (small insertions or deletions that lead to garbled protein downstream of the mutation) or large deletions. Mutations occurring near splice sites may cause exons to be skipped and important coding regions of the protein to be absent. All of these severe forms of mutations have been described in PPT deficient patients (Das et al., 1998; Mitchison et al., 1998; Munroe et al., 1998; Santorelli et al., 1998; Vesa et al., 1995; Waliany et al., 1999), with the exception of large deletions. Patients who inherit two such severe mutations will have no functional PPT from either allele; all of these patients studied to date have pursued a uniformly rapid and progressive course.

More subtle mutations (missense alleles) that cause amino acid substitutions may cause a range of enzyme activity, from very severe (no functional residual activity) to trivial, as in the case of amino acid polymorphisms that have only a slight effect on enzyme function. An example of a polymorphism in PPT is a rare isoleucine to threonine mutation that has been found in a parent of a child with INCL. In this case, one allele had a known INCL mutation and the polymorphism was found on the other allele. This polymorphism is found occasionally in the general population (1.5%) (Das et al., 1998). In other lysosomal storage disorders, "pseudo-deficiency" alleles have been described that result in even lower levels (10% or less) of enzyme activity as measured in a test tube, but which have no clinical significance (Hohenschutz et al., 1989; Nelson et al., 1991). Such alleles understandably create diagnostic confusion.

Other missense alleles have been found only in late-onset NCL patients. These mutations result in PPT enzymes with a small amount of residual activity. These mutations are referred to as late-infantile or juvenile NCL with GROD, to highlight the pathological finding that is characteristic of PPT deficiency. Typically, these patients are compound heterozygotes, with one severe allele (that contributes nothing) and one allele with a small amount of residual activity, in the range of 2–4% of normal.

B. Specific mutations in PPT: some examples

Over two dozen mutations occurring in all nine exons of PPT gene have been described (Das et al., 1998; Mitchison et al., 1998; Munroe et al., 1998; Santorelli et al., 1998; Vesa et al., 1995; Waliany et al., 1999). These mutations are illustrated in Figure 3.7 and include 13 missense, 6 nonsense, one single-base insertion, two single-base deletions, a 3-bp in-frame deletion, and a splice-site mutation. No larger deletions in the PPT gene have been reported. Certain mutations have occurred repeatedly or have been co-inherited with well-defined nonsense alleles that contribute no residual enzyme activity, allowing certain genotypes to be correlated with the age of onset of symptoms (Table 3.2).

In Finland, virtually all cases of infantile neuronal ceroid lipofuscinosis are caused by homozygous mutations at position 364 of the PPT cDNA, which changes a C to a T and causes a substitution of the amino acid arginine with a tryptophan residue at position 122 of the PPT protein (designated c.364C → T, Arg122Trp). These patients follow a uniformly rapid course, with first symptoms no later than 18 months and loss of higher cortical function by age 3. Brain tissue and other cells from Finnish NCL patients contains no detectable PPT protein, and when the allele is expressed in heterologous (COS) cells, the overexpressed protein localizes in the endoplasmic reticulum, the site of degradation of misfolded proteins. From these observations and from considerations of the effect of this amino acid change on the protein three-dimensional structure, it seems that the Arg122W mutations cause misfolding, destabilization, and rapid degradation of the mutant PPT protein.

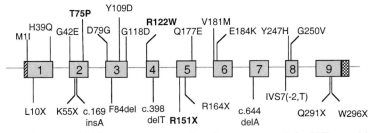

Figure 3.7. Location of 24 known mutations within the nine exons of the PPT gene. Missense mutations are shown above the gene and nonsense mutations, deletions and insertions are shown below. Common mutations are indicated in bold. Arg122Trp (R122W) is the common Finnish mutation and Arg151Stop (R151X) accounts for 40% of alleles outside of the Finnish population. Thr75Pro (T75P), also known as the "Scottish" mutation, accounts for most cases of juvenile-onset PPT deficiency worldwide.

Outside of Finland, the most common mutation is c.451C → T (Arg151Stop), which causes premature termination of the PPT protein (Das et al., 1998). Patients who are homozygous for this mutation have a course similar to Finnish infants. As one is unable to detect this mutation by PCR using cDNA from affected patients, it is assumed that this mutation is associated with low levels of mRNA. This observation is expected from the finding that nonsense mutations that occur at greater than 50 bp from 3′ donor splice sites cause message instability, due to coupling of transcription and translation by an unknown mechanism (Nagy and Maquat, 1998). The c. 451C → T mutation is in a position such that one would expect an unstable mRNA. No PPT protein is detected in these patients as well. Another nonsense mutation that has been observed in the homozygous state and that leads to a typical infantile phenotype is Leu10Stop (Munroe et al., 1998). Missense mutations that have occurred in typical infantile subjects include: c.541G → A (Val181Met), c.541G → T (Val181Leu) (Milà and Mallolas, 1998), c.125G → A (Gly42Glu), c.117T → A (His39Gln), c.550G → A (Glu184Lys), and c.353G → A (Gly118Asp). These mutations presumably do not contribute a significant amount of enzyme activity; however, some of these have been observed only once, and so more data are needed before firm conclusions regarding these mutations can be made.

One PPT mutation has been associated with a late-infantile phenotype in three families (Waliany et al., 1999). In each family, the age of onset of symptoms was 3–3.5 years. The mutation c.529C → G (Gln177Glu) would be predicted to cause a more subtle change in PPT structure. The defective enzyme has not been studied, but presumably has a small amount of residual enzymatic activity.

Several mutations in PPT have been associated with a juvenile onset NCL (so-called JNCL with GROD) (Das et al., 1998; Mitchison et al., 1998; Waliany et al., 1999). Typically, these children come to medical attention at the age of 5–7 years with visual difficulties. Parents report that color distortion is a common presenting finding, as is loss of central vision. A diagnosis of Leber's optic atrophy is sometimes made before more sophisticated tests are employed and before other signs appear. Unfortunately, cognitive decline and in some instances behavioral disturbances (including psychosis), become manifest from months to several years later, and progress to severe dementia and eventual death in the third to fourth decade. Anecdotally, seizures appear later and may be less severe as compared to patients with JNCL caused by mutations in CLN3.

The most common of the late-onset PPT mutations, observed in over a dozen families, is a c.223A → C (Thr75Pro) mutation. The Thr75Pro mutation occurs at a nonconserved site in PPT on a turn at the start of helix α1, which may disrupt the geometry of the α helix (Bellizzi et al., 1999). A small amount of residual activity associated with this mutant has been detected (Hofmann et al., 1999). Other mutations associated with juvenile-onset phenotypes are listed in Table 3.2. Interestingly, the mutations associated with JNCL with GROD are all

Table 3.2. Location of Mutations in the PPT Gene and Associated Phenotypes in Published Cases of NCL

Exon	Mutation	Amino acid change/effect	Phenotype	Number of alleles reported	References
1 (5′)	c.3G→A	MetIle	LINCL/JNCL[a]	2	(Das et al., 1998)
1 (5′)	c.29T→A	Leu10Stop	INCL	8	(Mitchison et al., 1998; Das et al., 1998; Munroe et al., 1998)
1 (3′)	c.117T→A	His39Gln	INCL	1	(Das et al., 1998)
2	c.125G→A	Gly42Glu	INCL	1	(Das et al., 1998)
2	c.163A→T	Lys55Stop	INCL	2	(Munroe et al., 1998; Vesa et al., 1995)
2	c.169-170insA	Frameshift	INCL	3	(Santorelli et al., 1998; Waliany et al., 1999)
2	c.223A→C	Thr75Pro	JNCL	18	(Das et al., 1998; Mitchison et al., 1998; Waliany et al., 1999)
3 (5′)	c.236A→G	Asp79Gly	JNCL[b]	1	(Mitchison et al., 1998)
3 (5′)	c.255-257delCTT	Phe84del	LINCL[b]	1	(Waliany et al., 1999)
3 (3′)	c.325T→G	Tyr109Asp	INCL/LINCL[a]	2	(Das et al., 1998)
3 (3′)	c.353G→A	Gly118Asp	INCL	1	(Waliany et al., 1999)
4	c.364A→T	Arg122Trp	INCL	85	(Das et al., 1998; Vesa et al., 1995)
4	c.398ΔT	Frameshift	INCL	1	(Das et al., 1998)
4	c.401T→C	Ile134Thr	Polymorphism	3	(Das et al., 1998)
5	c.451C→T	Arg151Stop	INCL	38	(Das et al., 1998; Mitchison et al., 1998; Munroe et al., 1998; Waliany et al., 1999)

5	c.490C→T	Arg164Stop	INCL	3	(Das et al., 1998; Waliany et al., 1999)
5	c.529C→G	Gln177Glu	INCL	3	(Das et al., 1998; Waliany et al., 1999)
6 (5′)	c.541G→A	Val181Met	INCL/LINCL/JNCL[c]	4	(Das et al., 1998; Waliany et al., 1999)
6 (5′)	c.550G→A	Glu184Lys	INCL	1	(Das et al., 1998)
7	c.644delA	Frameshift	—[d]	1	(Das et al., 1998)
7	c.656T→A	Leu219Gln	JNCL	1	(Mitchison et al., 1998)
8	IVS7-2A→T	Splice site	INCL	1	(Das et al., 1998)
8	c.739T→C	Tyr247His	JNCL[a]	1	(Das et al., 1998)
8	c.749G→T	Gly250Val	JNCL	1	(Das et al., 1998)
9 (5′)	c.871C→T	Gln291Stop	INCL	1	(Waliany et al., 1999)
9 (3′)	c.888G→A	Trp296Stop	—[e]	1	(Das et al., 1998)

[a]The M1I allele was co-inherited with either the Tyr247His and Tyr109Asp alleles in two subjects, a JNCL and LINCL subject, respectively. Therefore, the relative contribution of these three mutations to the phenotypes is unclear.

[b]The Phe84del allele was co-inherited with the Gln177Glu allele in a LINCL subject. Therefore, the relative contribution of this allele is unclear. However, Phe84 is an invariant residue throughout evolution, so Phe84del would be expected to be a severe mutation.

[c]The Val181Met allele has been co-inherited with Arg122Trp, Arg151Stop, and Thr75Pro in an INCL, LINCL, and two JNCL subjects, respectively. Therefore, the relative contribution of this allele is unclear. However, Val181 is an invariant residue throughout evolution, so the impact of the mutation would be expected to be severe.

[d]Unable to assess (co-inherited with Thr75Pro).

[e]Unable to assess (co-inherited with Gln177Glu).

expected to lead to small, local changes that only indirectly affect the catalytically active residues or fatty acid-binding site (Bellizzi et al., 1999).

X. LABORATORY DIAGNOSIS OF PPT DEFICIENCY

PPT deficiency may be suspected in any child with visual impairment, psychomotor deterioration, and seizures. In the very young child, microcephaly may be a presenting finding, and visual findings may be unappreciated. In the few PPT-deficient patients with presentations in the late-infantile age range (2–4 years), seizures have often been prominent, just as in classical LINCL. In older children, visual difficulties are usually the first manifestation, and anecdotal evidence suggests that seizures are not as prominent or occur later as compared with classical JNCL caused by mutations in CLN3. The finding of granular osmiophilic deposits on electron microscopic examination of peripheral blood leukocytes serves to narrow the differential diagnosis, although as many as 15% of INCL patients show nondiagnostic findings on initial examination (Wisniewski et al., 1992). Furthermore, late-onset PPT-deficient subjects often show a mixed picture consisting of GROD and curvilinear or fingerprint inclusions (Das et al., 1998). Therefore, caution must be used if ultrastructure alone is relied upon to determine the underlying enzyme deficiency, as these results may occasionally be misleading.

The most direct and precise means of confirming the diagnosis of PPT deficiency is based on the enzymatic assay of PPT from any available cell type (such as leukocytes or fibroblasts). A two-stage thioesterase assay utilizing a 4-methylumbelliferyl derivative of palmitoyl-S-thioglucose at acidic pH is available for clinical use (Voznyi et al., 1999). The substrate is available from Dr. Otto van Diggelen, Department of Clinical Genetics, Erasmus University, The Netherlands. DNA-based testing is applicable for the Finnish population, since the major mutation covers 98% of disease alleles (Vesa et al., 1995).

Prenatal testing has been applied successfully in Finland, by examination of chorionic villi, which show the characteristic GROD and by linkage studies for families with one affected child (Rapola et al., 1993). PPT activity can be measured in amniocytes (de Vries et al., 1999). DNA-based mutation detection can be offered to families seeking prenatal diagnosis for the disorder when the molecular defect in the family at risk has been established.

XI. PROSPECTS FOR CAUSE-SPECIFIC TREATMENT OF PPT DEFICIENCY

An appreciation that INCL and related disorders are caused by a deficiency in lysosomal PPT suggests two general approaches to therapy. One approach, applicable to all lysosomal storage diseases, seeks to replace the missing enzyme directly. Enzyme replacement therapy has proven to be remarkably effective in Gaucher disease, a lysosomal storage disease that affects bone and visceral organs (reviewed by Grabowski et al., 1998). However, enzyme replacement therapy for lysosomal storage disorders affecting the central nervous system faces the added problem of delivery to neuronal cells. The target organs in Gaucher disease are accessible from the bloodstream; in contrast, central nervous system neurons seem to be relatively inaccessible to blood-borne proteins, as they are protected by a blood–brain barrier as well as a cerebrospinal fluid–brain barrier. Some success in correcting central nervous system (CNS) disease has been observed in the case of the murine lysosomal storage disease (mucopolysaccharidosis VII) using treatment with the missing lysosomal enzyme, β-glucuronidase, early in postnatal life (O'Connor et al., 1998; Vogler et al., 1999). Little information on how enzymes may be safely delivered directly into the brain parenchyma is available, but preclinical animal studies are ongoing.

Generation of the missing enzyme within the brain through the use of viral-mediated gene delivery is an active area of investigation, but the field is still in its infancy. Obstacles to successful gene therapy in other systems have included difficulty in transfection of nondividing cells, low levels of expression, and development of antibodies to viral or (theoretically) to the therapeutic proteins. Some newer viral vectors show promise in overcoming some of these obstacles and are at the preclinical stage for selected lysosomal storage diseases (Barranger et al., 1999; Daly et al., 1999). Along these same lines, cells that produce the missing lysosomal enzyme might be used therapeutically. Neural progenitor stem cells have shown promise in a murine model of mucopolysaccharidosis VII (Snyder et al., 1995).

One way to introduce cells into the CNS is (surprisingly) through bone marrow transplantation. It seems that microglial cells of the CNS are bone-marrow derived, and slowly populate the brain after transplantation (Krivit et al., 1995b). The donor microglial cells (which are really a specialized form of tissue macrophage) secrete many lysosomal enzymes that are competent for uptake into neurons through the mannose 6-phosphate receptor pathway. Therefore, bone marrow transplantation provides a continuous cell-derived source of therapeutic enzyme. This approach has shown reasonable success when applied to some lysosomal storage diseases, but not to others (Hoogerbrugge and Valerio, 1998; Kaye, 1995; Krivit et al., 1995a, 1998; O'Marcaigh and Cowan, 1997). Four Finnish infants with INCL have received bone marrow transplants; one died of complications of the transplant, one has done much better than age-matched controls but is still severely retarded at age 4 years, and in the other two cases, follow-up has been too short to permit conclusions (P. Santavuori, personal communication). It is likely that bone marrow transplant may provide amelioration but will not be curative in these severely affected infants. The utility of bone marrow transplantion in juvenile-onset patients is unknown. In any event, an HLA-identical sibling (or matched anonymous donor) must be available, which further limits this approach. In addition, if an HLA-matched sibling is identified, the donor is likely to be a carrier of the disease mutation, and the procedure may (theoretically) be less successful because only half of the normal amount of enzyme will be available for correction.

New methods, such as correction of the defective gene in vivo by homologous recombination, using a method known as chimeraplasty, show some potential but have yet to be translated into the clinical arena (Xiang et al., 1997; Yoon et al., 1996). These newer technologies bear watching for the future.

A second conceptual approach to therapy for INCL would seek to deplete the abnormal substrate accumulation. There is a precedent to this approach in the glycosphingolipid disorders (Platt and Butters, 1998), in which a small molecule inhibitor of glycosphingolipid synthesis, N-butyldeoxynojirimycin, has been used to ameliorate lysosomal accumulations in mice with Tay-Sachs (Platt et al., 1997) and Sandhoff disease (Jeyakumar et al., 1999). Human clinical trials of the compound in Gaucher disease have been initiated. A second example of substrate depletion therapy is the use of cysteamine to treat cystinosis, a disorder of lysosomal cystine transport. In this instance, cysteamine is used to promote cystine breakdown and excretion through alternative transport mechanisms (Gahl et al., 1995).

In the case of INCL, little is known about the biosynthesis of lipid palmitoyl cysteine thioesters in cells, and no specific inhibitors are available. However, the concentration of accumulating lipid cysteine thioesters could possibly be reduced through the use of an appropriate reducing agent or nucleophile, because thioesters are high-energy, relatively unstable compounds amenable to chemical hydrolysis. An assay for the accumulation of cysteine thioesters such as the one presented in Figure 3.5 could be used to screen for such compounds. The use of pharmacological agents would be limited by their toxicity; for example, β-mercaptoethanol is very effective in breaking thioester bonds, but its reducing power makes it a very potent cellular poison. Perhaps less toxic compounds could be identified.

In summary, a great deal of progress has been made in our understanding of infantile NCL in recent years, offering hope that these advances will lead to better therapy for this devastating disease.

Acknowledgments

This work was supported in part by the National Institutes of Health (NS 36867 and NS 35323).

References

Barranger, J. A., Rice, E. O., and Swaney, W. P. (1999). Gene transfer approaches to the lysosomal storage disorders. *Neurochem. Res.* **24,** 601–615.
Bellizzi, J. J., III, Widom, J., Kemp, C., Lu, J.-Y., Das, A. K., Hofmann, S. L., and Clardy, J. (2000). The

crystal structure of palmitoyl protein thioesterase 1 and the molecular basis of infantile neuronal ceroid lipofuscinosis. *Proc. Natl. Acad. Sci. U.S.A.* **97,** 4573–4578.

Camp, L. A., and Hofmann, S. L. (1993). Purification and properties of a palmitoyl-protein thioesterase that cleaves palmitate from H-Ras. *J. Biol. Chem.* **268,** 22566–22574.

Camp, L. A., Verkruyse, L. A., Afendis, S. J., Slaughter, C. A., and Hofmann, S. L. (1994). Molecular cloning and expression of palmitoyl-protein thioesterase. *J. Biol. Chem.* **269,** 23212–23219.

Daly, T. M., Vogler, C., Levy, B., Haskins, M. E., and Sands, M. S. (1999). Neonatal gene transfer leads to widespread correction of pathology in a murine model of lysosomal storage disease. *Proc. Natl. Acad. Sci.* (USA) **96,** 2296–300.

Das, A. K., Becerra, C. H. R., Yi, W., Lu, J.-Y., Siakotos, A. N., Wisniewski, K. E., and Hofmann, S. L. (1998). Molecular genetics of palmitoyl-protein thioesterase deficiency in the U. S. *J. Clin. Invest.* **102,** 361–370.

Das, A. K., Bellizzi, J. J., III, Tandel, S., Biehl, E., Clardy, J., and Hofmann, S. L. (2000). Structural basis for the insensitivity of a serine enzyme (palmitoyl-protein thioesterase) to phenylmethylsulfonyl fluoride. *J. Biol. Chem.* **275,** 23847–23851.

de Vries, B. B., Kleijer, W. J., Keulemans, J. L., Voznyi, Y. V., Franken, P. F., Eurlings, M. C., Galjaard, R. J., Losekoot, M., Catsman-Berrevoets, C. E., Breuning, M. H., Taschner, P. E., and van Diggelen, O. P. (1999). First-trimester diagnosis of infantile neuronal ceroid lipofuscinosis (INCL) using PPT enzyme assay and CLN1 mutation analysis. *Prenat. Diagn.* **19,** 559–562.

Gahl, W. A., Schneider, J. A., and Aula, P. P. (1995). Lysosomal transport disorders: Cystinosis and sialic acid storage disorders In "The Metabolic and Molecular Bases of Inherited Disease" C. R. Scriver A. L. Beaudet, W. S. Sly and D. Valle, eds.), pp. 3763–3797. McGraw-Hill, New York.

Gold, A. M., and Fahrney, D. (1964). Sulfonyl fluorides as inhibitors of esterases. II. Formation and reactions of phenylmethanesulfonyl α-chymotrypsin *Biochemistry* **3,** 783–791.

Grabowski, G. A., Leslie, N., and Wenstrup, R. (1998). Enzyme therapy for Gaucher disease: The first 5 years. *Blood. Rev.* **12,** 115–33.

Haltia, M., Rapola, J., and Santavuori, P. (1973). Infantile type of so-called neuronal ceroid lipofuscinosis. Part II. Histological and electron microscopic studies. *Acta Neuropath.* **26,** 157–170.

Hellsten, E., Vesa, J., Heiskanen, M., Makela, T. P., Jarvela, I., Cowell, J. K., Mead, S., Alitalo, K., Palotie, A., and Peltonen, L. (1995). Identification of YAC clones for human chromosome 1p32 and physical mapping of the infantile neuronal ceroid lipofuscinosis (INCL) locus. *Genomics* **25,** 404–412.

Hellsten, E., Vesa, J., Olkkonen, V. M., Jalanko, A., and Peltonen, L. (1996). Human palmitoyl protein thioesterase: Evidence for lysosomal targeting of the enzyme and disturbed cellular routing in infantile neuronal ceroid lipofuscinosis. *Embo. J.* **15,** 5240–5245.

Hofmann, S. L., and Lu, J. Y. (1999). Metabolic labeling of protein-derived lipid thioesters in palmitoyl-protein thioesterase-deficient cells. *Meth. Mol. Biol.* **116,** 213–219.

Hofmann, S. L., and Verkruyse, L. A. (1999). Fatty acid analysis of protein-derived lipid thioesters isolated from palmitoyl-protein thioesterase-deficient cells. *Meth. Mol. Biol.* **116,** 221–228.

Hofmann, S. L., Das, A. K., Yi, W., Lu, J. Y., and Wisniewski, K. E. (1999). Genotype-phenotype correlations in neuronal ceroid lipofuscinosis due to palmitoyl-protein thioesterase deficiency. *Mol. Genet. Metab.* **66,** 234–239.

Hohenschutz, C., Eich, P., Friedl, W., Waheed, A., Conzelmann, E., and Propping, P. (1989). Pseudodeficiency of arylsulfatase A: A common genetic polymorphism with possible disease implications. *Hum. Genet.* **82,** 45–48.

Hoogerbrugge, P. M., and Valerio, D. (1998). Bone marrow transplantation and gene therapy for lysosomal storage diseases. *Bone Marrow Transplant.* **21 Suppl. 2,** S34–S36.

Jarvela, I. (1991). Infantile neuronal ceroid lipofuscinosis (CLN1): Linkage disequilibrium in the Finnish population and evidence that variant late infantile form (variant CLN2) represents a nonallelic locus. *Genomics* **10,** 333–337.

Jarvela, I., Schleutker, J., Haataja, L., Santavuori, P., Puhakka, L., and Manninen, T., et al. (1991). Infantile form of neuronal ceroid lipofuscinosis (CLN1) maps to the short arm of chromosome 1. *Genomics* **9**, 170–173.

Jeyakumar, M., Butters, T. D., Cortina-Borja, M., Hunnam, V., Proia, R. L., Perry, V. H., Dwek, R. A., and Platt, F. M. (1999). Delayed symptom onset and increased life expectancy in Sandhoff disease mice treated with N-butyldeoxynojirimycin. *Proc. Natl. Acad. Sci.* (USA) **96**, 6388–6393.

Kaye, E. M. (1995). Therapeutic approaches to lysosomal storage diseases. *Curr. Opin. Pediatr.* **7**, 650–654.

Kornfeld, S. (1990). Lysosomal enzyme targeting. *Biochem. Soc. Trans.* **18**, 367–374.

Krivit, W., Lockman, L. A., Watkins, P. A., Hirsch, J., and Shapiro, E. G. (1995a). The future for treatment by bone marrow transplantation for adrenoleukodystrophy, metachromatic leukodystrophy, globoid cell leukodystrophy and Hurler syndrome. *J. Inherit. Metab. Dis.* **18**, 398–412.

Krivit, W., Sung, J. H., Shapiro, E. G., and Lockman, L. A. (1995b). Microglia: The effector cell for reconstitution of the central nervous system following bone marrow transplantation for lysosomal and peroxisomal storage diseases. *Cell Transplant.* **4**, 385–392.

Krivit, W., Shapiro, E. G., Peters, C., Wagner, J. E., Cornu, G., Kurtzberg, J., Wenger, D. A., Kolodny, E. H., Vanier, M. T., Loes, D. J., Dusenbery, K., and Lockman, L. A. (1998). Hematopoietic stem-cell transplantation in globoid-cell leukodystrophy. *N. Engl. J. Med.* **338**, 1119–1126.

Lawson, D. M., Derewenda, U., Serre, L., Ferri, S., Szittner, R., Wei, Y., Meighen, E. A., and Derewenda, Z. S. (1994). Structure of a myristoyl-ACP-specific thioesterase from *Vibrio harveyi*. *Biochemistry* **33**, 9382–9388.

Lu, J. Y., Verkruyse, L. A., and Hofmann, S. L. (1996). Lipid thioesters derived from acylated proteins accumulate in infantile neuronal ceroid lipofuscinosis: Correction of the defect in lymphoblasts by recombinant palmitoyl-protein thioesterase. *Proc. Natl. Acad. Sci.* (USA) **93**, 10046–10050.

Milà, M., and Mallolas, J. (1998). Neuronal Ceroid Lipofuscinosis Database (Sara Mole, moderator), direct submission.

Mitchison, H. M., Hofmann, S. L., Becerra, C. H., Munroe, P. B., Lake, B. D., Crow, Y. J., Stephenson, J. B., Williams, R. E., Hofman, I. L., Taschner, P. E. M., Martin, J. J., Philippart, M., Andermann, E., Andermann, F., Mole, S. E., and Gardiner, R. M., and O'Rawe A. M. (1998). Mutations in the palmitoyl-protein thioesterase gene (PPT; CLN1) causing juvenile neuronal ceroid lipofuscinosis with granular osmiophilic deposits. *Hum. Mol. Genet.* **7**, 291–297.

Mumby, S. M. (1997). Reversible palmitoylation of signaling proteins. *Curr. Opin. Cell. Biol.* **9**, 148–154.

Munroe, P. B., Greene, N. D., Leung, K. Y., Mole, S. E., Gardiner, R. M., Mitchison, H. M., Stephenson, J. B., and Crow, Y. J. (1998). Sharing of PPT mutations between distinct clinical forms of neuronal ceroid lipofuscinoses in patients from Scotland [letter]. *J. Med. Genet.* **35**, 790.

Naggert, J., Witkowski, A., Mikkelsen, J., and Smith, S. (1988). Molecular cloning and sequencing of a cDNA encoding the thioesterase domain of the rat fatty acid synthetase. *J. Biol. Chem.* **263**, 1146–1150.

Nagy, E., and Maquat, L. E. (1998). A rule for termination-codon position within intron-containing genes: When nonsense affects RNA abundance. *Trends Biochem. Sci.* **23**, 198–199.

Nelson, P. V., Carey, W. F., and Morris, C. P. (1991). Population frequency of the arylsulphatase A pseudo-deficiency allele. *Hum. Genet.* **87**, 87–88.

O'Connor, L. H., Erway, L. C., Vogler, C. A., Sly, W. S., Nicholes, A., Grubb, J., Holmberg, S. W., Levy, B., and Sands, M. S. (1998). Enzyme replacement therapy for murine mucopolysaccharidosis type VII leads to improvements in behavior and auditory function. *J. Clin. Invest.* **101**, 1394–1400.

O'Marcaigh, A. S., and Cowan, M. J. (1997). Bone marrow transplantation for inherited diseases. *Curr. Opin. Oncol.* **9**, 126–130.

Platt, F. M., and Butters, T. D. (1998). New therapeutic prospects for the glycosphingolipid lysosomal storage diseases. *Biochem. Pharmacol.* **56**, 421–430.

Platt, F. M., Neises, G. R., Reinkensmeier, G., Townsend, M. J., Perry, V. H., Proia, R. L., Winchester, B., Dwek, R. A., and Butters, T. D. (1997). Prevention of lysosomal storage in Tay-Sachs mice treated with N-butyldeoxynojirimycin. *Science* **276**, 428–431.

Rapola, J., Salonen, R., Ammala, P., and Santavuori, P. (1993). Prenatal diagnosis of infantile neuronal ceroid-lipofuscinosis, INCL: Morphological aspects. *J. Inherit. Metab. Dis.* **16**, 349–352.

Santorelli, F. M., Bertini, E., Petruzzella, V., Di Capua, M., Calvieri, S., Gasparini, P., and Zeviani, M. (1998). A novel insertion mutation (A169i) in the CLN1 gene is associated with infantile neuronal ceroid lipofuscinosis in an Italian patient. *Biochem. Biophys. Res. Commun.* **245**, 519–522.

Schrag, J. D., and Cygler, M. (1997). Lipases and alpha/beta hydrolase fold. *Meth. Enzymol.* **284**, 85–107.

Schriner, J. E., Yi, W., and Hofmann, S. L. (1996). cDNA and genomic cloning of human palmitoyl-protein thioesterase (PPT), the enzyme defective in infantile neuronal ceroid lipofuscinosis [published erratum appears in *Genomics* (1996) **38**, 458]. *Genomics* **34**, 317–322.

Sleat, D. E., Sohar, I., Lackland, H., Majercak, J., and Lobel, P. (1996). Rat brain contains high levels of mannose-6-phosphorylated glycoproteins including lysosomal enzymes and palmitoyl-protein thioesterase, an enzyme implicated in infantile neuronal lipofuscinosis. *J. Biol. Chem.* **271**, 19191–19198.

Sleat, D. E., Kraus, S. R., Sohar, I., Lackland, H., and Lobel, P. (1997). alpha-Glucosidase and N-acetylglucosamine-6-sulphatase are the major mannose-6-phosphate glycoproteins in human urine. *Biochem. J.* **324**, 33–39.

Snyder, E. Y., Taylor, R. M., and Wolfe, J. H. (1995). Neural progenitor cell engraftment corrects lysosomal storage throughout the MPS VII mouse brain. *Nature* **374**, 367–370.

Soyombo, A. A., and Hofmann, S. L. (1997). Molecular cloning and expression of PPT2, a homolog of lysosomal palmitoyl-protein thioesterase with a distinct substrate specificity. *J. Biol. Chem.* **272**, 27456–27463.

van Diggelen, O. P., Keulemans, J. L., Winchester, B., Hofman, I. L., Vanhanen, S. L., Santavuori, P., and Voznyi, Y. V. (1999). A rapid fluorogenic palmitoyl-protein thioesterase assay: Pre- and postnatal diagnosis of INCL. *Mol. Genet. Metab.* **66**, 240–244.

Verkruyse, L. A., and Hofmann, S. L. (1996). Lysosomal targeting of palmitoyl-protein thioesterase. *J. Biol. Chem.* **271**, 15831–15836.

Verkruyse, L. A., Natowicz, M. R., and Hofmann, S. L. (1997). Palmitoyl-protein thioesterase deficiency in fibroblasts of individuals with infantile neuronal ceroid lipofuscinosis and I-cell disease. *Biochim. Biophys. Acta* **1361**, 1–5.

Vesa, J., Hellsten, E., Verkruyse, L. A., Camp, L. A., Rapola, J., Santavuori, P., Hofmann, S. L., and Peltonen, L. (1995). Mutations in the palmitoyl protein thioesterase gene causing infantile neuronal ceroid lipofuscinosis. *Nature* **376**, 584–587.

Vogler, C., Levy, B., Galvin, N. J., Thorpe, C., Sands, M. S., Barker, J. E., Baty, J., Birkenmeier, E. H., and Sly, W. S. (1999). Enzyme replacement in murine mucopolysaccharidosis type VII: Neuronal and glial response to beta-glucuronidase requires early initiation of enzyme replacement therapy. *Pediatr. Res.* **45**, 838–844.

Voznyi, Y. V., Keulemans, J. L. M., Mancini, G. M. S., Catsman-Berrevoets, C. E., Young, E., Winchester, B., Kleijer, W. J., and van Diggelen, O. P. (1999). A new simple enzyme assay for pre- and postnatal diagnosis of infantile neuronal ceroid lipofuscinosis (INCL) and its variants. *J. Med. Genet.* **36**, 471–474.

Waliany, S., Das, A. K., Gaben, A., Wisniewski, K. E., and Hofmann, S. L. (2000). Identification of three novel mutations of the palmitoyl-protein thioesterase-1 (*PPT1*) gene in children with neuronal ceroid-lipofuscinosis. *Hum. Mutat.* **15**, 206–207.

Wisniewski, K. E., Kida, E., Patxot, O. F., and Connell, F. (1992). Variability in the clinical and pathological findings in the neuronal ceroid lipofuscinoses: Review of data and observations. *Am. J. Med. Genet.* **42,** 525–532.

Witkowski, A., Witkowska, H. E., and Smith, S. (1994). Reengineering the specificity of a serine active-site enzyme: Two active-site mutations convert a hydrolase to a transferase. *J. Biol. Chem.* **269,** 379–383.

Xiang, Y., Cole-Strauss, A., Yoon, K., Gryn, J., and Kmiec, E. B. (1997). Targeted gene conversion in a mammalian CD34+-enriched cell population using a chimeric RNA/DNA oligonucleotide. *J. Mol. Med.* **75,** 829–835.

Yoon, K., Cole-Strauss, A., and Kmiec, E. B. (1996). Targeted gene correction of episomal DNA in mammalian cells mediated by a chimeric RNA.DNA oligonucleotide. *Proc. Natl. Acad. Sci. (USA)* **93,** 2071–2076.

4

Biochemistry of Neuronal Ceroid Lipofuscinoses

Mohammed A. Junaid* and Raju K. Pullarkat
Department of Developmental Biochemistry
New York State Institute for Basic Research in Developmental Disabilities
Staten Island, New York 10314

I. Introduction
II. Genetic Defects
   A. PPTI
   B. CLN2p
   C. CLN3 Protein
III. NCL are Lysosomal Storage Diseases
   A. Oligosaccharyl Diphosphodolichol
   B. Subunit c of Mitochondrial ATP Synthase
   C. Mannose 6-phosphorylated Glycoproteins
   D. Sphingolipid Activator Proteins (saposins)
IV. Remaining Issues and Future Directions
   References

ABSTRACT

This chapter summarizes the recent advances that have been made with respect to biochemical characterization of the neurodegenerative diseases collectively known as neuronal ceroid lipofuscinoses (NCL) or Batten disease. Genomic and proteomic approaches have presently identified eight different forms of NCL (namely, CLN1 through CLN8) based on mutations in specific genes. CLN1 and CLN2 are caused by mutations in genes that encodes lysosomal enzymes,

*Corresponding author: Institute for Basic Research, 1050 Forest Hill Road, Staten Island, NY 10314. Email: majunaid@aol.com.

palmitoyl protein thioesterase and pepstatin-insensitive proteinase, respectively. The protein involved in the etiology of CLN3 is a highly hydrophobic, presumably transmembrane protein.

NCL are considered as lysosomal storage diseases because of the accumulation of autofluorescent inclusion bodies. The composition of inclusion bodies varies in different forms of the NCL. The major storage component in CLN2 is the subunit c of mitochondrial ATP synthase complex and its accumulation is the direct result of lack of CLN2p in this disease. Mannose-6-phosphorylated glycoproteins accumulate in CLN3 and most likely their accumulation is the result of an intrinsic activity of the CLN3 protein. Significant levels of oligosaccharyl diphosphodolichol also accumulate in CLN3 and CLN2, whereas lysosomal sphingolipid activator proteins (saposins A and D) constitute major component of the storage material in CLN1.

The issue of selective loss of neuronal and retinal cells in NCL still remains to be addressed. Identification of natural substrates for the various enzymes involved in NCL may help in the characterization of the cytotoxic factor(s) and also in designing rationale therapeutic interventions for these group of devastating diseases.

I. INTRODUCTION

The term neuronal ceroid lipofuscinoses (NCL) or Batten disease refers to a group of familial, genetically heterogeneous neurodegenerative disorders that are characterized by the accumulation of autofluorescent inclusion bodies in brain and other tissues (Boustany, 1996; Zeman et al., 1970). Clinical characteristics of NCL include progressive visual failure, psychomotor deterioration, ataxia, seizures, and severe dementia that ultimately lead to premature death. Unfortunately, no effective treatment is available for this group of devastating neurodegenerative disorders, hence carrier identification by screening in the high-risk population and prenatal diagnosis are the only means of controlling NCL. The pathologic hallmark of NCL is the characteristic ultrastructural features of the accumulating autofluorescent inclusion bodies—a microscopic criterion that is used for diagnosis. Although systematic studies are lacking on the prevalence of NCL, these are considered to be the most common neurodegenerative diseases of infancy and childhood. Most of these disorders are inherited in an autosomal recessive manner, although there are at least two reports of families affected with the adult form of the disease that display autosomal dominant mode of inheritance (Dom et al., 1979; Boehme et al., 1971).

Earlier classifications of NCL into infantile, late-infantile, juvenile, and adult forms were based primarily on the age of onset of clinical symptoms and the characteristic ultrastructural appearance of the accumulated autofluorescent inclusion bodies. With the discovery of genetic defects, a new system of

Table 4.1. Genetic Variants of Neuronal Ceroid Lipofuscinoses

Trivial name	Disease	Onset	Pathology	Locus	Product	Reference
Infantile						
Infantile	CLN1	6–18 mo	GROD	1p32	PPTI	Vesa et al., 1995
Late-infantile						
Classic	CLN2	2–4 yr	CB	11p15	CLN2p/TPP1	Sleat et al., 1997
Finnish variant	CLN5	4–7 yr	CB/FP	13p22	Membrane protein	Savukoski et al., 1998
Costa Rican variant	CLN6	4–8 yr	CB/FP	15q21-23	Unknown	Sharp et al., 1997
Turkish variant	CLN7	4–8 yr	CB/FP	?	Unknown	Wheeler et al., 1999
Juvenile						
Classical	CLN3	6–10 yr	FP	16p12	Membrane protein	International Batten Disease Consortium, 1995
EPMR/NR	CLN8	5–10 yr	RP	8p23	Membrane protein	Ranta et al., 1999
Adult						
Kufs disease	CLN4	~30 yr	G/FP/RP	?	?	Berkovic et al., 1988

GROD, granular osmiophilic deposits; CB, curvilinear bodies; FP, fingerprint profiles; RP, rectilinear profiles; EPMR/NR, progressive epilepsy with mental retardation or Northern epilepsy; PPTI, palmitoyl-protein thioesterase; TPP1, tripeptidyl peptidase 1.

classification has been introduced that identifies the disease with its gene. Currently, at least eight different forms of NCL have been identified based on the genetic defects (see Table 4.1).

II. GENETIC DEFECTS

In the past few years, advances in positional cloning and proteomics have led to the chromosomal mapping and gene identification of seven of the eight NCL. Only the gene for CLN4 remains to be identified. Two of the seven gene products have already been purified and characterized. These are for the diseases CLN1 and CLN2, caused by mutations in genes that encode lysosomal enzymes, palmitoyl-protein thioesterase (PPTI) (Vesa et al., 1995) and pepstatin-insensitive proteinase (CLN2p) (Sleat et al., 1997), respectively. Genetic defects and chromosomal localization for two relatively rare late-infantile forms, CLN5 (Savukoski et al., 1998) and CLN6 (Sharp et al., 1997) have been identified. Among these, the gene product of CLN5 has been proposed as a 407-amino acid residues long, hypothetical membrane-bound protein based on the predicted hydrophobicity index. While a new locus has been identified for another rare form, CLN7, the gene and the product remains to be identified (Wheeler et al., 1999). The majority of CLN3 patients carry a 1.02-kb deletion in the gene that encodes a putative highly hydrophobic transmembrane protein (International Batten Disease Consortium, 1995). This cln3 gene is proposed to transcribe a protein of 438 amino acid residues. Most recently, another disease, referred to as progressive epilepsy with mental retardation or Northern epilepsy (EPMR/NR), which is restricted mostly to the Finnish population, has been included under NCL. The genetic defect for this disease, categorized as CLN8, has been identified (Ranta et al., 1999). The gene is predicted to transcribe a putative highly hydrophobic transmembrane protein of 286 amino acid residues. A naturally occurring animal model for NCL known as motor neuron degeneration (mnd) mouse also has mutations in an orthologous gene for cln8.

A. PPTI

PPTI is a soluble lysosomal thioesterase involved in the removal of fatty acyl moieties from proteins (Verkruyse and Hoffman, 1996). The enzyme occurs as a doublet of molecular mass 39 and 37 kDa, presumably as a result of differential glycosylation, and has a pH optimum between 4 and 5. Unlike other lipid modifications such as N-terminal myristoylation or C-terminal cysteine prenylation, PPTI removes fatty acyl moieties from thioester linkages. Lymphoid cells derived from CLN1 patients accumulate [^{35}S] cysteine-labeled lipid thioesters—a defect that can be reversed by including recombinant PPTI in the culture medium (Lu et al., 1996). The identities of these accumulating lipid thioesters are still unknown. Under in vitro conditions, depalmitoylation of H-Ras and the $G_{\alpha s}$ subunits of heterotrimeric GTP-binding proteins by the PPTI have been demonstrated, suggesting that these could be natural substrates (Camp and Hofmann, 1993). However, storage of H-Ras or any peptide fragments originating from it in CLN1 is not known. Data from our laboratory has shown that the levels of H-Ras remain unaffected in CLN1 (Morris and Pullarkat, unpublished). The natural substrate(s) of the PPTI, thus, remains to be identified. The molecular pathogenesis of CLN1 presumably results from defective intracellular transport of the enzymatically inactive PPTI due to a single point mutation that causes a change of a tryptophan to an arginine residue. In transfection experiments utilizing COS-1 cells, it was demonstrated that while the wild-type PPTI is sorted to lysosomes, the mutant inactive protein carrying the same mutation as that occurring in CLN1 patients remains in the endoplasmic reticulum (Hellsten et al., 1996).

Biochemical diagnosis of CLN1 that utilize either radiolabeled (Cho and Dawson, 1998) or fluorogenic (van Diggelen et al., 1999) palmitoylated peptides as substrates for PPTI assay have been described. With these assays, detection of patients as well as heterozygous carriers is possible.

B. CLN2p

CLN2p is one of the most abundant lysosomal proteases and is insensitive to the common aspartyl protease inhibitor pepstatin A. When CLN2p was discovered as a missing mannose 6-phosphorylated protein in CLN2 patients, it was reported to be a novel protein (Sleat et al., 1997). However, later GenBank sequence submission revealed it to be the same tripeptidyl peptidase 1 (TPPI) that was earlier purified from rat spleen (Rawlings and Barrett, 1999; Vines and Warburton, 1998, 1999). The amino acid sequence of CLN2p is highly conserved among various species. The murine sequence is about 84% identical to the human sequence (GenBank accession numbers AJ011912 and AF111172).

The CLN2p/TPPI is transcribed as a single polypeptide of molecular mass 61 kDa that is proteolytically cleaved and posttranslationally modified to the mature single-polypeptide, 46-kDa enzyme (Junaid et al., 2000). The mature protein has five N-glycosylation sites, and all are probably glycosylated. The enzyme cleaves tripeptide from amino-termini of peptides that bear free α-amino groups (Junaid et al., 2000; Vines and Warburton, 1998). It was claimed that the rat spleen TPPI also possesses a carboxypeptidase activity against angiotensin II (Vines and Warburton, 1998). In contrast, the CLN2p purified from bovine or human brain was absolutely free of any carboxypeptidase activity (Junaid et al., 2000). Moreover, unlike the TPPI, the CLN2p cleave tripeptides sequentially only upon prolonged incubation (Junaid and Pullarkat, unpublished). Under in vitro conditions, CLN2p proteolytically cleave naturally occurring neuropeptides such as angiotensin II, substance P and β-amyloid (Junaid et al., 2000). The significance of CLN2p/TPPI activity against β-amyloid is presently unclear, since, in Alzheimer disease patients, the activity of this proteinase is found to be significantly elevated in brain (Junaid and Pullarkat, 1999). The purified CLN2p has a broad substrate specificity under in vitro conditions and cleave peptide bonds adjacent to neutral, acidic, and basic residues. The crucial natural substrate for CLN2p still remains to be identified.

A highly specific biochemical diagnostic test for CLN2 based on the measurement of CLN2p/TPPI activity in patients and heterozygous carriers has been described (Junaid et al., 1999). This test is based on the CLN2p/TPPI activity against a synthetic tetrapeptide linked to 7-amino-4-(trifluoromethyl)coumarin. The diagnostic assay can be performed in leukocytes prepared from heparinized blood samples (Table 4.2). There is no detectable CLN2p/TPPI activity in red blood corpuscles. Two other assays for CLN2p/TPPI for the diagnosis of CLN2 have also been described (Sohar et al., 1999; Vines and Warburton, 1999), but these assays detect residual peptidase activities in CLN2 patients.

Table 4.2. CLN2p/TPPI Activity in Human Blood

	Number	CLN2p/TPPI activity[a] nmol/h/mg protein	Range
RBC		Not detected	
Leukocytes			
Normal controls	11	1947 ± 153	1633–2107
CLN2 patients	10	Not detected	
Obligate heterozygotes	38	1002 ± 220	442–1287

[a]Enzyme activity is expressed as average mean ±SD.

While CLN2p/TPPI is deficient in CLN2, its activity is increased in several other neurodegenerative disorders (Junaid and Pullarkat, 1999). This increased CLN2p activity may result from infiltrating glial cells, since, all these neurodegenerative diseases result in neuronal loss followed by glial cell proliferation. In neurological conditions such as Krabbe disease or autism, where there is no evidence of any significant neuronal loss, the CLN2p/TPPI activity remains unaffected.

C. CLN3 protein

Despite rigorous attempts by a number of laboratories, the identity of the CLN3 protein remains elusive. The predicted translation product is a 483-residue protein with four N-glycosylation sites (International Batten Disease Consortium, 1995). The putative CLN3 protein was also shown to undergo phosphorylation both at serine and threonine residues (Michalewski et al., 1998). A prenylation motif exists at the C-terminus, and there is in vitro evidence for farnesylation of a cysteine residue at this motif (Pullarkat and Morris, 1997). Based on its hydrophobicity index, the CLN3 protein has been proposed to have six transmembrane segments (Janes et al., 1996). This model however, does not account for the prenylation motif, and the farnesylated C-terminus is also expected to be embedded in the cytoplasmic side of the membrane. The presence of a prenyl group in CLN3 protein may offer a unique situation whereby a transmembrane protein possesses a farnesyl membrane anchor.

The absence of a good antibody is the main reason for lack of functional characterization of the CLN3 protein. Mostly, the antibodies for CLN3 protein were generated against short synthetic peptide sequences. These antibodies have provided conflicting results with regard to the molecular mass and cellular localization of the CLN3 protein. Antibodies generated against peptides comprising of either residues 4–19 and 242–258 showed the presence of 43-kDa protein in cells stably expressing the CLN3 protein (Jarvela et al., 1998, 1999).

Based on the open reading frame of the *cln3* gene, a molecular mass of 48 kDa is predicted without any post-translational modifications. There is no evidence where the truncation of the nascent CLN3 polypeptide is occurring in the expressed protein. The cellular localization of the CLN3 protein also remains disputed with claims of localization in lysosomes (Jarvela *et al.*, 1998), Golgi bodies (Kremmidiotis *et al.*, 1999), and mitochondria (Katz *et al.*, 1997). These studies show the CLN3 protein as 43-kDa and 50-kDa proteins in transfected cells and mouse tissues, respectively. In primary neuronal cultures, the protein was found to be concentrated in the synaptic ends and growth cones. A yeast ortholog *btn1* for the *cln3* gene was found to be localized in vacuoles that are the equivalent of lysosomes (Croopnick *et al.*, 1998). Functionally, in yeast the *btn1* product was found to be nonessential for cell viability, mitochondrial function, or mitochondrial protein catabolism (Pearce and Sherman, 1997). It was, however, found to be associated with regulation of the lysosomal pH (Pearce *et al.*, 1999b). Yeast lacking the *btn1* gene were found to have an abnormally acidic vacuole, and this condition could be reversed by including chloroquine in the culture medium (Pearce *et al.*, 1999a).

Recently, two laboratories have reported the generation of knockout mice for CLN3 by targeted disruption of the orthologous gene (Mitchison *et al.*, 1999; Katz *et al.*, 1999). These mice accumulate autofluorescent inclusion bodies in various cell types—a characteristic feature found in CLN3 patients. Loss of certain cortical interneurons as well as hypertrophy of hippocampal interneurons was clearly evident in the knockout mouse. In addition, higher CLN2p activity was observed in the mouse brain, consistent with that observed in autopsy brain samples of CLN3 patients. These knockout mice should greatly facilitate deciphering of the functional aspects of the CLN3 protein.

III. NCL ARE LYSOSOMAL STORAGE DISEASES

NCL have long been considered as storage diseases because of the accumulation of autofluorescent inclusion bodies in various cell types. Evidence is now emerging that implicate NCL as a group of lysosomal storage diseases. Characterization of the inclusion bodies have revealed it to be a complex mixture of lipids and proteins, and their composition varies in different forms of the disease.

A. Oligosaccharyl diphosphodolichol

Lipids isolated from NCL brain samples were found to contain elevated levels of dolichol, which is also excreted in urine (Hall and Patrick, 1985; Wolfe *et al.*, 1986). Several investigators have shown the storage of oligosaccharyl diphosphodolichol (oligo-PP-dol) in brain tissues from patients (Pullarkat et al., 1988; Hall and Patrick, 1988). The increase in oligo-PP-dol was more pronounced in CLN2 and CLN3, while CLN1 showed only marginal change. The accumulating oligo-PP-dol contains di-N-acetylglucosamine residues attached to tetra- or penta-mannosyl moiety that in turn is phosphorylated (Hall and Patrick, 1988). The elevated dolichols appear to be secondary biochemical effects rather than a primary defect in NCL. Several other neurodegenerative diseases, such as Down syndrome, Alzheimer, Huntington, and Creutzfeld-Jacob, also display elevated urinary excretion of dolichols.

B. Subunit c of mitochondrial ATP synthase

Initial extensive studies with ovine model for NCL and later with CLN2 have demonstrated that the major storage protein in these diseases is the highly hydrophobic subunit c of mitochondrial ATP synthase (Palmer et al., 1986, 1992). A specific delay in the degradation of subunit c due to cellular proteolytic dysfunction was shown in the fibroblasts derived from CLN2 patients (Ezaki et al., 1996). A lysosomal pepstatin-insensitive proteinase (CLN2p) was later found to be the underlying genetic defect in CLN2 (Sleat et al., 1997). Synthetic amino-terminal peptide fragments from subunit c sequence are substrates for the CLN2p (Junaid et al., 2000). These peptide fragments can be proteolytically cleaved by fibroblast extracts from normal subjects, while extracts from CLN2 patients lack this proteolytic activity. This clearly demonstrates that the accumulation of subunit c is the direct result of lack of CLN2p. It is unclear, however, why subunit c accumulates in other NCL forms such as CLN3 or *mnd* mouse, which has mutation in the gene orthologous for *cln8*, in spite of severalfold higher CLN2p activity being present (Junaid and Pullarkat, 1999). It is also interesting that subunit c accumulates in other unrelated lysosomal diseases such as mucopolysaccharidoses type I-III, mucolipidosis type I, GM1 and GM2 gangliosidoses, and Niemann-Pick disease types A and C (Elleder et al., 1997). Thus the significance of subunit c in the pathophysiology of NCL is unclear.

While CLN2p cleaves synthetic amino-terminal peptides of varying lengths from subunit c sequence, the native peptide isolated from CLN2 patients could not be hydrolyzed. On the other hand, purified CLN2p can cleave purified subunit c from bovine brain. It is possible that the accumulated subunit c in CLN2 patients undergo secondary modification near the N-terminal, thereby becoming inaccessible for CLN2p proteolysis. We have observed that CLN2p cleaves only those peptides that have free amino-terminal residue, whereas peptides with blocked amino groups of the N-terminal residue are not substrates. There are reports of ϵ-N-trimethylation of the lysine^{43} of subunit c in canine NCL model (Katz et al., 1994). However, such a modification so far from the proteinase cleavage site is not expected to exert any effect.

C. Mannose 6-phosphorylated glycoproteins

The posttranslational modification in the form of mannose 6-phosphate is generated on newly synthesized lysosomal proteins (Kornfeld, 1987). It serves as a recognition signal for mannose 6-phosphate receptors of proteins that are destined for lysosomes, wherein the recognition signal is believed to be enzymatically removed. Initially, the levels of a 46-kDa glycoprotein was found to be drastically elevated in CLN3, while at the same time it was missing in CLN2 (Pullarkat and Zawitosky, 1993). This 46-kDa protein was later identified as the deficient mannose 6-phosphorylated CLN2p that resulted from the underlying genetic defect in this disease. In CLN3, the overall levels of all mannose 6-phosphorylated proteins are drastically elevated, while in CLN4, two of these proteins are increased (Sleat et al., 1998). The activities of various lysosomal enzymes, including CLN2p, were also found to be elevated in CLN3, but the increase in CLN2p was not proportional to the increase in the mannose 6-phosphorylated form. It was found that at least mannose 6-phosphorylated form of CLN2p is enzymatically inactive (Junaid and Pullarkat, 1999). This might be necessary for a proteinase to remain in an inactive form while it is being transported to its destination.

D. Sphingolipid activator proteins (saposins)

Saposins are small-molecular-mass, heat-stable lysosomal proteins that originate from a single precursor molecule that cleaves into four different peptides. These proteins activate lysosomal hydrolases involved in the degradation of sphingolipids. The major storage materials in the CLN1 are two of the saposins A and D (Tyynela et al., 1993). Accumulation of these two proteins seems to be an indirect effect, since saposins accumulation have also been demonstrated in CLN3, Gaucher, Niemann-Pick types A and B, fucosidosis, Tay-Sachs, and infantile Sandhoff diseases. Thus, the significance of saposins accumulation in NCL is unclear.

IV. REMAINING ISSUES AND FUTURE DIRECTIONS

The remarkable progress in the past few years in identifying the genetic defects in seven of eight forms of NCL is indeed a success story. Since accumulation of autofluorescent inclusion bodies in lysosomes is the pathological hallmark of NCL, it is tempting to speculate that these group of disorders may have common metabolic aberrations. Chemical analyses of the isolated inclusion bodies have demonstrated the heterogeneous nature of the compounds among different forms of NCL. Also, the identity of compound(s) causing autofluorescence is still unknown. Neuronal death followed by gliosis is another common feature of NCL. What causes the neuronal and retinal cell degeneration in NCL needs to be addressed. The selective neuronal and retinal loss most likely results from accumulation of specific factors that may be peptides. The fact that the symptoms manifest several years after birth, as evident in CLN3 and CLN4, implies that accumulation of these factors above certain threshold levels are necessary to elicit the toxicity. It is possible that materials accumulating in lysosomes due to deficiencies of specific proteins are spilled out into the extracellular milieu and cause cytotoxicity. In lysosomal storage diseases it is not uncommon to detect storage materials in urine (Wisniewski et al., 1995) and cerebrospinal fluid (Pullarkat et al., 1981). Identification of these cellular degenerative factors and their pharmacological depletion will certainly help in designing rationale therapeutic intervention for these devastating diseases.

A protease deficiency as the underlying defect in CLN2 is clearly established. Are other forms of NCL also caused by protease deficiency? PPTI is involved in the hydrolysis of fatty acyl thioesters. It may also be possible that PPTI itself has proteolytic activity, since a number of esterases and thioesterases also possess protease activity (Crews et al., 1996; Cho and Cronan, 1994). This may be the reason why saposins constitute the major portion of the accumulated inclusion bodies in CLN1. The CLN2p/TPPI is a peptidase most likely involved in the degradation of subunit c, but it is less likely that the neurodegeneration is related to the accumulating subunit c. It is also unlikely that cells synthesize one proteinase to degrade only one peptide/protein. Thus, other peptides or proteins that are natural substrates of the CLN2p/TPPI remain to be identified. Marked storage of mannose 6-phosphorylated glycoproteins in CLN3 suggest that the CLN3 protein may be a phosphatase involved in the removal of phosphate groups from lysosomally targeted mannose 6-phosphorylated proteins. Similar to the glucose 6-phosphatase system, the CLN3 protein as a membrane-bound entity may possess multifunctional activities involved in the transport of lysosomal proteins into the organelle as well as dephosphorylation of the mannose 6-phosphate moiety. So far, the identity of the phosphatase involved in dephosphorylation of mannose 6-phosphate moiety from lysosomally targeted glycoproteins has remained elusive despite extensive search (Bresciani and von Figura, 1996; Bresciani et al., 1992). Also, contrary to expectation, activities of certain abnormal acid phosphatases are elevated in CLN3 (Faisal Khan et al., 1995). There is no evidence of any genetic redundancy in CLN3 that could account for elevated acid phosphatase activity. A situation similar to the lack of activity of mannose 6-phosphorylated form of CLN2 may also be possible in CLN3, where another lysosomal proteinase remains inactive due to lack of dephosphorylation. In this context it is interesting to mention that urine samples from CLN3 patients were found to contain significant amounts of low-molecular-mass peptides (LaBadie and Pullarkat, 1990).

Acknowledgments

Supported in part by a grant from the National Institutes of Health (NS 30147) and funds from New York State Office of Mental Retardation and Developmental Disabilities.

References

Berkovic, S. F., Carpenter, S., Andermann, F., Andermann, E., and Wolfe, L. S. (1988). Kufs disease: A critical reappraisal. *Brain* **111,** 27–67.
Boehme, D. H., Cottrell, J. C., Leonberg, S. C., and Zeman, W. (1971). A dominant form of neuronal ceroid lipofuscinosis. *Brain* **94,** 745–760.
Boustany, R.-M. (1996). Neurodystrophies and neurolipidoses. In "Handbook of Clinical Neurology" (H. W. Moser, ed.), Vol. 22 (66), pp. 671–700. Elsevier Science, Amsterdam, The Netherlands.
Bresciani, R., and von Figura, K. (1996). Dephosphorylation of the mannose-6-phosphate recognition marker is localized in late compartments of the endocytic route: Identification of purple acid phosphate (uteroferrin) as the candidate phosphatase. *Eur. J. Biochem.* **238,** 669–674.
Bresciani, R., Peters, C., and von Figura, K. (1992). Lysosomal acid phosphatase is not involved in the dephosphorylation of mannose 6-phosphate containing lysosomal proteins. *Eur. J. Cell Biol.* **58,** 57–61.
Camp, L. A., and Hoffman, S. L. (1993). Purification and properties of a palmitoyl protein thioesterase that cleaves palmitate from H-Ras. *J. Biol. Chem.* **268,** 22566–22574.
Cho, H., and Cronan, J. E. (1994). Protease I of *Escherichia coli* functions as a thioesterase *in vivo. J. Bacteriol.* **176,** 1793–1795.
Cho, S., and Dawson, G. (1998). Enzymatic and molecular biological analysis of palmitoyl protein thioesterase deficiency in infantile neuronal ceroid lipofuscinosis. *J. Neurochem.* **71,** 323–329.
Crews, C. M., Lane, W. S., and Schreiber, S. L. (1996). Didemnin binds to the protein palmitoyl thioesterase responsible for infantile neuronal ceroid lipofuscinosis. *Proc. Natl. Acad. Sci. U.S.A.* **93,** 4316–4319.
Croopnick, J. B., Choi, H. C., and Mueller, D. M. (1998). The subcellular location of the yeast *Saccharomyces cerevisiae* homologue of the protein defective in the juvenile form of Batten disease. *Biochem. Biophys. Res. Commun.* **250,** 335–341.
Dom, R., Brucher, J. M., Ceuterick, C., Carlton, H., and Martin, J. J. (1979). Adult ceroid lipofuscinosis (Kufs disease) in two brothers: Retinal and visceral storage in one; diagnostic muscle biopsy in the other. *Acta Neuropathol.* **45,** 67–72.
Elleder, M., Sokolova, J., and Hrebicek, M. (1997). Follow-up study of subunit c of mitochondrial ATP synthase (SCMAS) in Batten disease and in unrelated lysosomal disorders. *Acta Neuropathol.* **93,** 379–390.
Ezaki, J., Wolfe, L. S., and Kominami, E. (1996). Specific delay in the degradation of mitochondrial ATP synthase subunit c in late infantile neuronal ceroid lipofuscinosis is derived from cellular proteolytic dysfunction rather than structural alteration of subunit c. *J. Neurochem.* **67,** 1677–1687.
Faisal Khan, K. M., Sklower Brooks, S., and Pullarkat, R. K. (1995). Abnormal acid phosphatases in neuronal ceroid-lipofuscinoses. *Am. J. Med. Genet.* **57,** 285–289.
Hall, N. A., and Patrick, A. D. (1985). Dolichol and phosphorylated dolichol content of tissues in ceroid-lipofuscinosis. *J. Inher. Metab. Dis.* **8,** 178–183.
Hall, N. A., and Patrick, A. D. (1988). Accumulation of dolichol-linked oligosaccharides in ceroid-lipofuscinosis (Batten disease). *Am. J. Med. Genet.* **Suppl. 5,** 221–232.
Hellsten, E., Vesa, J., Olkkonen, V. M., Jalanko, A., and Peltonen, L. (1996). Human palmitoyl

protein thioesterase: Evidence for lysosomal targeting of the enzyme and disturbed cellular routing in infantile neuronal ceroid lipofuscinosis. *EMBO J.* **15,** 5240–5245.

International Batten Disease Consortium (1995). Isolation of a novel gene underlying Batten disease, CLN3. *Cell* **82,** 949–957.

Janes, R. W., Munroe, P. B., Mitchison, H. M., Gardiner, R. M., Mole, S. E., and Wallace, B. A. (1996). A model for Batten disease protein CLN3: Functional implications from homology and mutations. *FEBS Lett.* **399,** 75–77.

Jarvela, I., Sainio, M., Rantamaki, T., Olkkonen, V. M., Carpen, O., Peltonen, L., and Jalanko, A. (1998). Biosynthesis and intracellular targeting of the CLN3 protein defective in Batten disease. *Hum. Mol. Genet.* **7,** 85–90.

Jarvela, I., Lehtovirta, M., Tikkanen, R., Kyttala, A., and Jalanko, A. (1999). Defective intracellular transport of CLN3 is the molecular basis of Batten disease (JNCL). *Hum. Mol. Genet.* **8,** 1091–1098.

Junaid, M. A., and Pullarkat, R. K. (1999). Increased brain lysosomal pepstatin-insensitive proteinase activity in patients with neurodegenerative diseases. *Neurosci. Lett.* **264,** 157–160.

Junaid, M. A., Sklower Brooks, S., Wisniewski, K. E., and Pullarkat, R. K. (1999). A novel assay for lysosomal pepstatin-insensitive proteinase and its application for the diagnosis of late-infantile neuronal ceroid lipofuscinosis. *Clin. Chim. Acta* **281,** 169–176.

Junaid, M. A., Wu, G., and Pullarkat, R. K. (2000). Purification and characterization of bovine brain lysosomal pepstatin-insensitive proteinase, the gene product deficient in the human late-infantile neuronal ceroid lipofuscinosis. *J. Neurochem.* **74,** 287–294.

Katz, M. L., Christianson, J. S., Norbury, N. E., Gao, C.-L., Siakotos, A. N., and Koppang, N. (1994). Lysine methylation of mitochondrial ATP synthase subunit c stored in tissues of dogs with hereditary ceroid lipofuscinosis. *J. Biol. Chem.* **269,** 9906–9911.

Katz, M. L., Gao, C.-L., Prabhakaram, M., Shibuya, H., Liu, P.-C., and Johnson, G. S. (1997). Immunochemical localization of the Batten disease (CLN3) protein in retina. *Invest. Opthalmol. Vis. Sci.* **38,** 2375–2386.

Katz, M. L., Shibuya, H., Liu, P.-C., Kaur, S., Gao, C.-L., and Johnson, G. S. (1999). A mouse gene knockout model for juvenile ceroid-lipofuscinosis (Batten disease). *J. Neurosci. Res.* **57,** 551–556.

Kornfeld, S. (1987). The trafficking of lysosomal enzymes. *FASEB J.* **1,** 462–468.

Kremmidiotis, G., Lensink, I. L., Bilton, R. L., Woollatt, E., Chataway, T. K., Sutherland, G. R., and Callen, D. F. (1999). The Batten disease gene product (CLN3p) is a golgi integral membrane protein. *Hum. Mol. Genet.* **8,** 523–531.

LaBadie, G. U., and Pullarkat, R. K. (1990). Low molecular weight urinary peptides in ceroid lipofuscinoses: Potential biochemical markers for the juvenile subtype. *Am. J. Med. Genet.* **37,** 592–599.

Lu, J.-Y., Verkruyse, L. A., and Hofmann, S. L. (1996). Lipid thioesters derived from acylated proteins accumulate in infantile ceroid lipofuscinosis: Correction of the defect in lymphoblasts by recombinant palmitoyl-protein thioesterase. *Proc. Natl. Acad. Sci. (USA)* **93,** 1046–1050.

Michalewski, M. P., Kaczmarski, W., Golabek, A. A., Kida, E., Kaczmarski, A., and Wisniewski, K. E. (1998). Evidence for phosphorylation of CLN3 protein associated with Batten disease. *Biochem. Biophys. Res. Commun.* **253,** 458–462.

Mitchison, H. M., Bernard, D. J., Greene, N. D. E., Cooper, J. D., Junaid, M. A., Pullarkat, R. K., de Vos, N., Breuning, M. H., Owens, J. W., Mobley, W. C., Gardiner, R. M., Lake, B. D., Taschner, P. E. M., and Nussbaum, R. L. (1999). Targeted disruption of the *cln3* gene provides a mouse model for batten disease. *Neurobiol. Dis.* **6,** 321–334.

Palmer, D. N., Barns, G., Husbands, D. R., and Jolly, R. D. (1986). Ceroid lipofuscinosis in sheep: II. The major component of the lipopigment in liver, kidney, pancreas, and brain is low molecular weight protein. *J. Biol. Chem.* **261,** 1773–1777.

Palmer, D. N., Fearnley, I. M., Walker, J. E., Hall, N. A., Lake, B. D., Wolfe, L. S., Haltia, M., Martinus, R. D., and Jolly, R. D. (1992). Mitochondrial ATP synthase subunit c storage in the ceroid-lipofuscinoses (Batten disease). *Am. J. Med. Genet.* **42,** 561–567.

Pearce, D. A., and Sherman, F. (1997). *BTN1*, a yeast gene corresponding to the human gene responsible for Batten's disease, is not essential for viability, mitochondrial function, or degradation of mitochondrial ATP synthase. *Yeast* **13,** 691–697.

Pearce, D. A., Carr, C. J., Das, B., and Sherman, F. (1999a). Phenotypic reversal of the *btn1* defects in yeast by chloroquine: A yeast model for Batten disease. *Proc. Natl. Acad. Sci. (USA)* **96,** 11341–11345.

Pearce, D. A., Ferea, T., Nosel, S. A., Das, B., and Sherman, F. (1999b). Action of *BTN1*, the yeast orthologue of the gene mutated in Batten disease. *Nat. Genet.* **22,** 55–58.

Pullarkat, R. K., and Morris, G. N. (1997). Farnesylation of Batten disease CLN3 protein. *Neuropediatrics* **28,** 42–44.

Pullarkat, R. K., and Zawitosky, S. E. (1993). Glycoconjugate abnormalities in the ceroid-lipofuscinoses. *J. Inher. Metab. Dis.* **16,** 317–322.

Pullarkat, R. K., Reha, H., and Beratis, N. G. (1981). Accumulation of ganglioside GM2 in cerebrospinal fluid of a patient with the variant AB of infantile GM2 gangliosidosis. *Pediatrics* **68,** 106–108.

Pullarkat, R. K., Kim, K. S., Sklower, S. L., and Patel, V. K. (1988). Oligosaccharyl diphosphodolichols in the ceroid-lipofuscinoses. *Am. J. Med. Genet.* **Suppl. 5,** 243–251.

Ranta, S., Zhang, Y., Ross, B., Lonka, L., Takkunen, E., Messer, A., Sharp, J., Wheeler, R., Kusumi, K., Mole, S., Liu, W., Soares, M. B., Bonaldo, M., de, F., Hirvasniemi, A., de la Chapelle, A., Gilliam, T. C., and Lehesjoki, A.-E. (1999). The neuronal ceroid lipofuscinoses in human EPMR and *mnd* mutant mice are associated with mutations in *CLN8*. *Nat. Genet.* **23,** 233–236.

Rawlings, N. D., and Barrett, A. J. (1999). Tripeptidyl-peptidase I is apparently the CLN2 protein absent in classical late-infantile neuronal ceroid lipofuscinosis. *Biochem. Biophys. Acta* **1429,** 496–500.

Savukoski, M., Klockars, T., Holmberg, V., Santavuori, P., Lander, E. S., and Peltonen, L. (1998). CLN5, a novel gene encoding a putative transmembrane protein mutated in Finnish variant late infantile neuronal ceroid lipofuscinosis. *Nat. Genet.* **19,** 286–288.

Sharp, J. D., Wheeler, R. B., Lake, B. D., Savukowski, M., Jarvela, I. E., Peltonen, L., Gardiner, R. M., and Williams, R. E. (1997). Loci for classical and a variant late infantile neuronal ceroid lipofuscinoses map to chromosome 11p15 and 15q21-23. *Hum. Mol. Genet.* **6,** 591–596.

Sleat, D. E., Donnelly, R. J., Lackland, H., Liu, C.-G., Sohar, I., Pullarkat, R. K., and Lobel, P. (1997). Association of mutations in a lysosomal protein with classical late-infantile neuronal ceroid lipofuscinosis. *Science* **277,** 1802–1805.

Sleat, D. E., Sohar, I., Pullarkat, P. S., Lobel, P., and Pullarkat, R. K. (1998). Specific alterations in levels of mannose 6-phosphorylated glycoproteins in different neuronal ceroid lipofuscinoses. *Biochem. J.* **334,** 547–551.

Sohar, I., Sleat, D. E., Jadot, M., and Lobel, P. (1999). Biochemical characterization of a lysosomal protease deficient in classical late infantile neuronal ceroid lipofuscinosis (LINCL) and development of an enzyme-based assay for diagnosis and exclusion of LINCL in human specimens and animal models. *J. Neurochem.* **73,** 700–711.

Tyynela, J., Palmer, D. N., Baumann, M., and Haltia, M. (1993). Storage of saposins A and D in infantile neuronal ceroid-lipofuscinosis. *Fed. Eur. Biochem. Soc.* **330,** 8–12.

van Diggelen, O. P., Keulemans, J. L. M., Winchester, B., Hofman, I. L., Vanhanen, S. L., Santavuori, P., and Voznyi, Y. V. (1999). A rapid fluorogenic palmitoyl-protein thioesterase assay: Pre- and postnatal diagnosis of INCL. *Mol. Genet. Metab.* **66,** 240–244.

Verkruyse, L. A., and Hofmann, S. L. (1996). Lysosomal targeting of palmitoyl-protein thioesterase. *J. Biol. Chem.* **271,** 15831–15836.

Vesa, J., Hellsten, E., Verkruyse, L. A., Camp, L. A., Rapola, J., Santavuori, P., Hofmann, S. L., and Peltonen, L. (1995). Mutations in the palmitoyl protein thioesterase gene causing infantile neuronal ceroid lipofuscinosis. *Nature* **376**, 584–587.

Vines, D. J., and Warburton, M. J. (1998). Purification and characterization of a tripeptidyl aminopeptidase I from rat spleen. *Biochim. Biophys. Acta* **1384**, 233–242.

Vines, D. J., and Warburton, M. J. (1999). Classical late infantile neuronal ceroid lipofuscinosis fibroblasts are deficient in lysosomal tripeptidyl peptidase I. *FEBS Lett.* **443**, 131–135.

Wheeler, R. B., Sharp, J. D., Mitchell, W. A., Bate, S. L., Williams, R. E., Lake, B. D., and Gardiner, R. M. (1999). A new locus for variant late infantile neuronal ceroid lipofuscinosis—CLN7. *Mol. Genet. Metab.* **66**, 337–338.

Wisniewski, K. E., Kaczmarski, W., Golabek, A. A., and Kida, E. (1995). Rapid detection of subunit c of mitochondrial ATP synthase in urine as a diagnostic screening method for neuronal ceroid-lipofuscinoses. *Am. J. Med. Genet* **57**, 246–249.

Wolfe, L. S., Palo, J., Santavuori, P., Andermann, F., Andermann, E., Jacob, J. C., and Kolodny, E. (1986). Urinary sediment dolichols in the diagnosis of neuronal ceroid-lipofuscinosis. *Ann. Neurol.* **19**, 270–274.

Zeman, W., Donahue, S., Dyken, P., and Green, J. (1970). Leukodystrophies and poliodystrophies. In "Handbook of Clinical Neurology" (P. J. Vinken, and G. W. Bruyn, eds.), Vol. 10, pp. 588–679. Elsevier Science Amsterdam, The Netherlands.

5

Positional Cloning of the JNCL Gene, *CLN3*

Terry J. Lerner*
Molecular Neurogenetics Unit
Massachusetts General Hospital
Charlestown, Massachusetts 02129
and
Department of Neurology
Harvard Medical School
Boston, Massachusetts 02114

I. Introduction
II. *CLN3* Maps to Chromosome 16
III. A Subunit 9 Gene Is Not *CLN3*
IV. Refined Localization of *CLN3*
V. Physical Mapping of the *CLN3* Candidate Region
VI. Exon Trapping Yields a Candidate cDNA
VII. The Common Mutation in JNCL is a 1-kb Genomic Deletion
VIII. Mutational Analysis of the *CLN3* Gene
IX. Tissue Expression of *CLN3*
X. *CLN3* Encodes a Novel Protein
XI. Animal Models of JNCL
References

I. INTRODUCTION

Juvenile neuronal ceroid lipofuscinosis (*CLN3*, JNCL, Batten or Spielmeyer-Vogt-Sjogren disease, MIM304200) is the most common neurodegenerative disorder of childhood. Inheritance of the disease is autosomal recessive. Its incidence

*Address for correspondence: E-mail: Lerner@helix.mgh.harvard.edu

is estimated at 1–5/100,00 births (Zeman, 1974), with an increased prevalence in the northern European population. Onset begins with visual failure between the ages of 5 and 10 years. Seizures and mental deterioration manifest several years later, leading to death in the second or third decade. There is no cure for JNCL. Treatment is largely symptomatic.

Diagnostic criteria of JNCL include neurological studies (electroencephalograms, visual-evoked responses, and electroretinograms); magnetic resonance imaging (MRI); elevated urinary dolichol levels; and electron microscopic examination of lymphocytes and skin, conjunctiva, or rectal tissues. Peripheral blood lymphocytes appear vacuolated. Deep skin and rectal biopsies show striking deposits of a PAS-positive, sudanophilic, autofluorescent storage material. These inclusions are filled with membranes that appear as "fingerprint profiles" on electron microscopy (Santavuori, 1988; Wisniewski et al., 1988). With the isolation and characterization of the JNCL gene *CLN3* in 1995 (International Batten Disease Consortium, 1995), direct, reliable DNA diagnosis of JNCL has became possible for the majority of patients (see Sections VII and VIII).

In 1993, the International Batten Disease Consortium was founded to accelerate the search for *CLN3*. Members included the laboratories of Dr. Terry Lerner (Massachusetts General Hospital), Dr. R. Mark Gardiner (University College London Medical School), Dr. Martijn Beuning (Leiden University), Dr. David Callen (Adelaide Women's and Children's Hospital), and Dr. Norman Doggett (Los Alamos National Laboratory). As a result of this collaboration, researchers in each laboratory gained access to an international patient resource, detailed genetic maps based on highly informative microsatellite markers, extensive genome-wide and chromosome-specific physical maps, and the most efficient tools for isolating genes.

II. *CLN3* MAPS TO CHROMOSOME 16

Eiberg (1989) first mapped the JNCL locus *CLN3* to chromosome 16. Linkage was confirmed with the classical protein marker haptoglobin on the long arm of the chromosome (16q22) with a LOD score of 3 at $\Theta = .00$ in males and $\Theta = .26$ in females. Subsequent multipoint analysis with DNA RFLP markers confirmed linkage to the chromosome and mapped the gene to a 15-cM region, flanked by *D16S150* and *D16S148* and spanning the centromere (Gardiner et al., 1990). In a later study, the same laboratory reported a new localization of the gene in a 2.3-cM region between *D16S148* and *D16S67*, outside of the original locus and on the short arm (p) of the chromosome (Callen et al., 1991). Using the marker *D16S285* (Konradi et al., 1991), Yan et al. (1993) showed that *CLN3* does not localize to this region in the U.S. collection of JNCL families and mapped the gene back to the original candidate region between *D16S148* and *D16S150*.

III. A SUBUNIT 9 GENE IS NOT *CLN3*

Palmer (1992) demonstrated the abnormal accumulation of the proteolipid subunit 9 in the inclusions found in both JNCL and LINCL, where it accounts for 19% and 85% of the protein mass, respectively. This protein is a normal component of H^+-ATP synthase, a multisubunit ATP-generating enzyme at the end of the mitochondrial respiratory chain. Subunit 9 transports protons across the inner mitochondrial membrane (Sebald and Hoppe, 1981).

In many genetic human diseases involving the accumulation of an insouble protein, cloning of the gene for that protein has led directly to the genetic defect. Experiments mapping the subunit 9 genes *P1* and *P2* to chromosomes 17 and 12, respectively, excluded these genes as the site of *CLN3* (Dyer and Walker, 1993). Yan *et al.* (1994) discovered a previously unreported subunit 9 gene, *P3*, and demonstrated the mapping of this gene to chromosome 2, also excluding it as the JNCL gene.

IV. REFINED LOCALIZATION OF *CLN3*

Refined localization of *CLN3* only became possible with the advent of highly polymorphic microsatellite markers. Linkage studies by two independent laboratories (Mitchison *et al.*, 1993; Lerner *et al.*, 1994) yielded significant evidence for strong allelic association (linkage disequilibrium) between *CLN3* and the microsatellite markers at *D16S288* (Shen *et al.*, 1991), *D16S298* and *D16S299* (Thompson *et al.*, 1992), and also the RFLP marker *D16S272* (Lerner *et al.*, 1992, 1994). The finding of linkage disequilibrium between JNCL and the markers in this region is consistent with a founder effect in the disease.

A collaborative study was initiated to analyze the pooled family resource of these two laboratories with the goal of fine-mapping the gene (Mitchison *et al.*, 1994). One hundred forty-two JNCL families with a total of 200 affected and 161 unaffected siblings and originating from 16 different countries were included. Five markers, *D16S288*, *D16S299*, *D16S298*, *SPN* (Rogaev and Keryanov, 1992), and *D16S383* (Shen *et al.*, 1993), were typed in all families.

Crossover events were observed for both *D16S288* and *D16S383*, defining new flanking markers (Figure 5.1). No crossover events were observed between the gene and *D16S298*, *D16S299*, and *SPN*. These results localize the gene to a 2.1-cM (sex-averaged) interval on 16p12.1-p11.1 between the markers *D16S288* and *D16S383*.

Within the *D16S288-D16S383* interval, four microsatellite markers (*D16S288*, *D16S299*, *D16S298*, and *SPN*) are in strong disequilibrium with *CLN3* (Table 5.1). A high-risk haplotype is defined based on the 7 allele (142 bp) at *D16S288*, the 5 allele (118 bp) at *D16S299*, the 6 allele (180 bp) at *D16S298*,

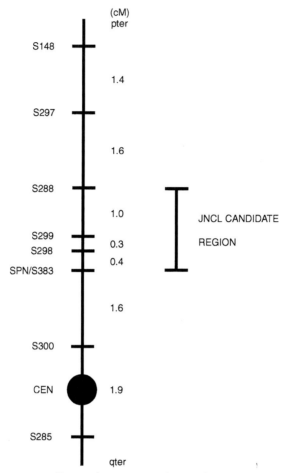

Figure 5.1. A genetic map of human chromosome 16, showing the DNA markers used in the refined localization of the JNCL gene *CLN3*. Distances are in centimorgans.

and the 1 allele (251 bp) at *SPN*. The "7561" haplotype is found on 28% of disease chromosomes and none of the normal chromosomes.

A core haplotype with the 5 allele at *D16S299* and the 6 allele at *D16S298* is greatly enriched, present on 73% of the disease chromosomes and on only 3% of the control nondisease chromosomes. Recombination between these two markers is rare. The haplotype 75** is never found without the associated 6 allele at *D16S298* on the disease chromosome, whereas the haplotype **61 is present on 7% of disease chromosomes and 8% of control chromosomes. These results, taken together, indicate that ancient recombinations between *D16S299* and *D16S288* and between *D16S298* and *SPN* appear to have been relatively frequent, while

Table 5.1. Distribution of Haplotype Frequencies on Disease and Control Chromosomes

Haplotype				Disease chromosomes		Control chromosomes	
S288	S299	S298	SPN	No.	Frequency	No.	Frequency
7	5	6	1	72	0.28	0	0.00
*	5	6	1	68	0.26	2	0.01
7	5	6	*	29	0.11	0	0.00
*	5	6	*	22	0.08	5	0.02
*	*	6	1	17	0.07	22	0.08
7	5	*	*	0	0.00	3	0.01
Other disease haplotypes				52	0.20	14	0.05
Other control haplotypes				0	0.00	228	0.83
Total				260	1.00	274	1.00

recombination events between $D16S299$ and $D16S298$ have been rare. These results strongly suggest that $CLN3$ lies closest to $D16S299$ and $D16S298$.

This conclusion is further supported by genetic studies of the isolated Finnish population (Mitchison et al., 1995). The 6 allele at $D16S298$ is present on 96% of Finnish JNCL chromosomes. Estimates using the Luria-Delbruck equation predict that $CLN3$ lies 8.8 kb (range 6.3–13.8) from $D16S298$ and 165.4 kb (range 132.4–218.1 kb) from $D16S299$. Thus, it is likely that the major mutation found in Finland is the same as that found in other European countries and the United States.

In addition, two patients were identified who have deletions of the $D16S298$ locus. One patient, who is of Moroccan origin and from a consanguineous relationship, is homozygous for the null allele at $D16S298$ (Taschner et al., 1995). The extent of the deletion in this patient was estimated by PCR and hybridization analysis to be ~29 kb, with 5 kb distal to and ~24 kb proximal to the $D16S298$ locus. A second patient from Finland was found to be hemizygous for $D16S298$ (International Batten Disease Consortium, 1995). These independent findings support the linkage disequilibrium mapping results defining $D16S298$ as the marker closest $CLN3$.

V. PHYSICAL MAPPING OF THE *CLN3* CANDIDATE REGION

In parallel to efforts to refine the genetic locus of $CLN3$, a physical map consisting of yeast artificial chromosomes (YACs) and cosmid contigs was generated to span the candidate region (Jarvela et al., 1996). Oligonucleotide primers at each $CLN3$-related locus were used to screen the CEPH YAC libraries (Albertson et al., 1990; Cohen et al., 1993) and a flow-sorted chromosome 16 YAC library (McCormick et al., 1993). The resultant map consists of 72 YACs and encompasses the closest flanking markers (*D16S288* and *D16S383*), including both *D16S299* and *D16S298*. The map was ordered using 42 sequence-tagged sites (STS): tel-*D16S297-IL4R-D16S288-D16S272-D16S299-D16S298-D16S48-SPN/D16S383*.

The YAC clones, which had been selected with primers for the *D16S298* locus, proved unstable, when sized by pulsed field gel electrophoresis and analyzed for STS content. Therefore, chromosome walking was initiated in the Los Alamos flow-sorted chromosome 16 cosmid library (Stallings et al., 1992) and a total human genomic cosmid library (Blonden et al., 1989) to obtain a cosmid contig mapping to the deletion in the Moroccan patient. The walk, starting from the *D16S48* locus in Los Alamos contig c343.1 and proceeding toward the *D16S299* locus, yielded a 185-kb contig cNL/343.1 (Taschner et al., 1995). Cosmid NL11A, the most distal clone in the contig, was found to contain the *D16S298* locus and to encompass the region deleted in the Moroccan patient (Figure 5.2).

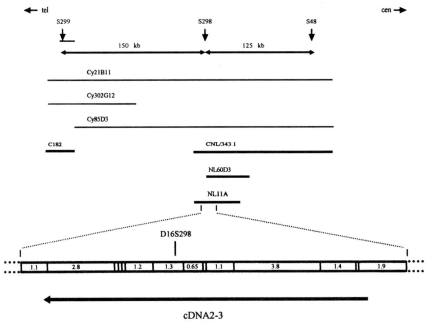

Figure 5.2. A physical map of the *CLN3* candidate region (International Batten Disease Consortium, 1995). The positions of selected DNA markers used for linkage and haplotype analyses are indicated. Individual cosmids NL11A and NL60D3 of cosmid contig CNL/343.1, which contains *D16S298* and *D16S48*, and cosmid contig C182, which contains *D16S299* are shown. Three yeast artificial chromosomes (Cy21B11, Cy30G12, and Cy85D3) that form part of a 980-kb contig spanning the candidate region are indicated by horizontal lines. PstI restriction fragments of NL11A that could be ordered are shown. The genomic extent of cDNA2-3 is shown below the map, with the arrow indicating the direction of transcription. (This figure has been reproduced with the permission of Cell Press.)

VI. EXON TRAPPING YIELDS A CANDIDATE cDNA

Cosmid clone NL11A, the most likely site of the JNCL gene *CLN3*, was selected for exon trapping for the isolation of candidate JNCL gene transcripts. Exon amplification was carried out using the pSPL3 vector and PstI-digested NL11A target DNA (Church et al., 1994; Lerner et al., 1995). A 18-bp exon was isolated and used to screen a commercial fetal brain cDNA library (Stratagene), yielding a 1.7-kb cDNA (clone 2-3). Mapping studies showed that clone 2-3 is contained in its entirety within cosmid NL11A, spans the *D16S298* locus, and overlaps the deletion found in the Moroccan patient, thus making it an optimal candidate for *CLN3*.

VII. THE COMMON MUTATION IN JNCL IS A 1-kb GENOMIC DELETION

To screen for possible genomic alterations associated with the cDNA2-3 gene, Southern blots of PstI-digested DNA from unrelated JNCL patients were scanned by hybridization with a probe, representing either the 5′ or the 3′ half of the cDNA (Figure 5.3A). When the 5′ probe was used, patients homozygous for the *D16S299/D16S298* "56" haplotype (lanes 1–3) displayed the loss of a 3.8-kb PstI fragment, also present in controls (lane 4), and gained a novel 2.8-kb fragment. Patients heterozygous for the *D16S299/D16S298* "56" haplotype displayed both 3.8- and 2.8-kb bands (lane 5). The restriction patterns were also altered in patients with the "56" haplotype when several other enzymes were tested, supporting the hypothesis of a genomic deletion present on the "56" chromosome. Figure 5.3B demonstrates the Mendelian inheritance of this deletion in a family with an affected child homozygous for the "56" haplotype.

To determine the effect of this genomic deletion on the cDNA transcript, PCR amplification of RT-cDNA from patients homozygous for the "56" chromosome haplotype was carried out. Sequence analysis of the rt-PCR products revealed the loss of 217 bp of internal coding sequence from the full-length control cDNA (see Section X). This deletion causes a frameshift, generating a novel termination codon 84 bp downstream of the deletion junction. The predicted product is a truncated protein of 181 amino acids, compared to 438 amino acids in the control.

Further testing confirmed the 1-kb genomic deletion of the clone 2-3 gene associated with the *D16S299/D16S298* "56" haplotype as the most common mutation in JNCL, accounting for at least 81% of disease chromosomes. Eighty-one unrelated JNCL patients representing 24 haplotypes and originating from 16 countries were screened for the presence of the 1-kb deletion (Table 5.2). Of these patients, 46 were homozygous for the "56" haplotype, 24 were heterozygous for

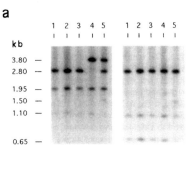

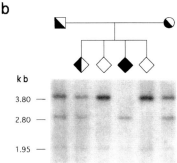

Figure 5.3. (a) Southern blot analysis of cDNA2-3 in JNCL patients (International Batten Disease Consortium, 1995). Genomic DNA was digested with PstI, fractionated by agarose gel electrophoresis, and transferred to nylon membrane. Lanes 1–3, patients homozygous for the "56" alleles; lane 4, unaffected control; lane 5, patient heterozygous for the "56" alleles. The left panel was probed with a radiolabeled PCR fragment corresponding to the 5′ half of cDNA2-3. The right panel was probed with a radiolabeled PCR fragment corresponding to the 3′ half of cDNA2-3. (b) Southern blot analysis of cDNA2-3 in a family with an affected child homozygous for the "56" haplotype. Southern blots were prepared as above and hybridized with the 5′ probe. The affected child is shown by the closed symbol and the carriers by the half-closed symbols. Progeny are shown by diamonds, and the birth order of some individuals has been changed to protect confidentiality. The chromosomes segregating in this pedigree have been distinguished by extensive typing with markers mapping to the candidate region (Lerner et al., 1994). (This figure has been reproduced with the permission of Cell Press.)

the "56" haplotype, and 11 did not carry the "56" haplotype on either chromosome. In all 70 patients with an affected "56" chromosome, the deleted fragment was detected and in all 46 homozygotes for this haplotype, only the deleted fragment was found. Smaller numbers of chromosomes bearing closely related haplotypes also carried this deletion, suggesting that these chromosomes most probably derived from the 56 chromosome by mutation of the polymorphic marker or recombination.

Table 5.2. Deletion Analysis of JNCL Patients

D16S299/D16S298 haplotype	2.5-kb fragment[a]	3.5-kb fragment[a]	Number of chromosomes[b]
56	+	−	116
56	−	+	0
Related	+	−	15
Related	−	+	8
Other	+	−	1
Other	−	+	22

[a]Genomic PCR was carried out using a primer pair flanking the deletion, yielding a 3.5-kb amplification product on control chromosomes and a 2.5-kb amplification product on chromosomes bearing the 1-kb deletion.
[b]The total number of chromosomes was 162.

VIII. MUTATIONAL ANALYSIS OF THE *CLN3* GENE

The finding of additional mutations of the gene encoding cDNA2-3 in JNCL patients supported the conclusion that this gene is the JNCL gene *CLN3*. Subsequent mutational analysis revealed, in addition to the common 1-kb deletion and the deletions found in the Moroccan and Finnish patient, 20 mutations in 11 of the 15 exons of the 2-3 gene (International Batten Disease Consortium, 1995; Mitchison et al., 1997; Munroe et al., 1997). No mutations were found in the gene sequence upstream of exon 5. While the majority of these novel mutations are private, i.e., found in only one family, five of the mutations occur in more than one family. Furthermore, all families with the same mutation have the same or closely related haplotypes, suggesting the presence of smaller founder effects.

The 20 mutations include 6 missense mutations, 5 nonsense mutations, 3 small deletions, 3 small insertions, 2 splice-site mutations, and one intronic mutation (Figure 5.4). The majority of patients have mutations that are predicted to give rise to truncated proteins, and these patients all have the classical JNCL phenotype. Several patients, heterozygous for the 1-kb deletion and a missense mutation, have nonclassical phenotypes. One patient, with the leu101pro mutation, remained conversant until death at the age of 25. Two other patients, heterozygous for the 1-kb deletion and either the leu170Pro or the Glu295lys mutation, have a disease that is dominated by visual failure. Both these patients have nearly normal IQ and completed higher education.

Phenotypic variability within families with more than one affected child is also observed. For example, the affected older sibling of one of the patients with the predominant visual phenotype died at the age of 22 with the classical disease phenotype. Intrafamilial variation is seen in many diseases, but its cause is unknown. Genetic and environmental factors, undoubtedly, come into play.

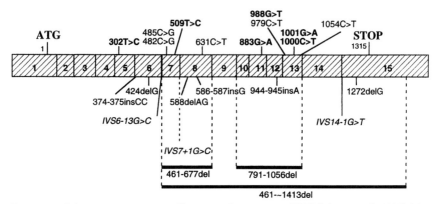

Figure 5.4. Schematic representation of locations of mutations in CLN3 (Munroe et al., 1997). Mutations identified above the cDNA are point mutations in the open reading frame and mutations below the cDNA are deletions, insertions, or point mutations in introns. Missense mutations are in boldface and intronic mutations are in italics. Large genomic deletions are shown as solid horizontal bars. The deleted nucleotides relate to the cDNA only. (This figure has been reproduced with the permission of The University of Chicago Press.)

Because of the slow progression of symptoms in JNCL and its similarity to other neurological disorders, diagnosis is often missed or delayed. Now, with the identification of the mutations underlying this disease, direct DNA testing is available for the majority of families for the diagnosis and prenatal and carrier testing of JNCL (Taschner et al., 1997).

IX. TISSUE EXPRESSION OF *CLN3*

Northern blot analysis, using the CLN3 cDNA as probe, showed a 1.7-kb transcript in mRNA isolated from the brain, the site of massive neuronal death in JNCL patients, and from a wide variety of tissue (International Batten Disease Consortium, 1995). Consistent with these findings, inclusions are found in many JNCL tissues in addition to the brain. The tissue expression of CLN3 reflects the expression pattern of the other NCL genes so far identified (Vesa et al., 1995; Sleat et al., 1997; Savukoski et al., 1998), and offers no clue toward understanding the neuronal specificity of this group of disorders.

X. *CLN3* ENCODES A NOVEL PROTEIN

Sequence analysis shows that the CLN3 cDNA spans 1689 bp, including a predicted open reading frame of 1314 bp (International Batten Disease Consortium,

1995; GeneBank Accession Number U32680). The predicted product is a protein of 438 amino acids with a molecular mass of 48 kDa. No homology has been found at either the nucleotide or the protein level with any known gene, indicating that CLN3 encodes a novel protein of unknown function.

Computer predictions for the CLN3 polypeptide reveal matches for N-glycosylation sites, glycosaminoglycan attachment sites, c-AMP- and cGMP-dependent protein kinase phosphorylation sites, protein kinase C phosphorylation sites, casein kinase II phosphorylation sites, and myristoylation sites. The PSORT program for prediction of protein localization sites (Nakai and Kanehisa, 1992) indicates that the CLN3 protein may be a membrane protein. This localization is supported by hydropathy calculations that suggest the presence of several hydrophobic domains and by the numerous potential N-glycosylation and myristoylation sites.

XI. ANIMAL MODELS OF JNCL

Homologs of the CLN3 protein have been identified in numerous species, as diverse as yeast (Genebank Accession Number Z49334) and the mouse (Genebank Accession Number U47106; Lee et al., 1996). The protein product of the yeast gene *BTN1* is 39% identical and 59% similar to the human CLN3 protein. The product of the mouse gene *Cln3* is 85% identical to the human protein (Lee et al., 1996). The conservation of this protein across species suggests that it plays an important role in the cell. The missense mutations so far identified in humans affect amino acid residues that are conserved in these homologs, indicating that these residues must be critical for the functions of the CLN3 protein.

Studies of a yeast model implicate a role for the *CLN3* protein in the regulation of lysosomal pH (Pearce and Sherman, 1997, 1998; Pearce et al., 1999). Overexpression studies localize the BTN1 protein to the yeast vacuole, the equivalent of the mammalian lysosome (Croopnick et al., 1998). Yeast knockouts, lacking the *BTN1* gene, have abnormally acidic vacuoles in the early stages of growth. It is postulated that this defect in the regulation of lysosomal pH may lead to aggregation of lysosomal proteins, including the *CLN1* and *CLN2* gene products, or to altered activity of lysosomal proteases.

Two mouse strains, the *mnd* and the *nclf*, also develop a disease that closely resembles human NCL (Bronson et al., 1993, 1998). However, neither is caused by mutations in *Cln3*, the mouse locus that lies on mouse chromosome 7 (Lee et al., 1996). The *mnd* mutation maps to a gene on mouse chromosome 8 (Messer et al., 1992) that is the homolog of a newly recognized subtype of NCL, *CLN8*, the gene for progressive epilepsy with mental retardation (Ranta et al., 1999). The *nclf* mutation maps to mouse chromosome 9 (Bronson et al., 1998) and is syntenic with *CLN6* (Sharp et al., 1997; Haines et al., 1998). In the absence of a natural mouse model for JNCL, several groups have used gene-targeting techniques to generate mice bearing mutations in *Cln3* (Greene et al., 1999; Katz et al., 1999; Lerner et al., unpublished). These mice develop the inclusions that are the hallmark of the disease and exhibit some neuropathological abnormalities (Mitchison et al., 1999). These animal models will be critical for studies of the pathogenesis underlying the disease and for the evaluation of therapies.

Acknowledgments

The author would like to thank the JNCL families and their physicians. Without their help and encouragement, the identification of *CLN3* would not have been possible. TJL is funded by grants from the National Institute of Neurological Disorders and Stroke (NS32099), The JNCL Research Fund, The Children's Brain Diseases Foundation, and The Batten Disease Support and Research Association.

References

Albertson, H. M., Abderrahim, H., Cann, H. M., Dausset, J., Le Paslier, D., and Cohen, D. (1990). Construction and characterization of a yeast artificial chromosome library containing seven haploid human genome equivalents. *Proc. Natl. Acad. Sci. (USA)* **87,** 4256–4260.

Blonden, L. A. J., den Dunnen, J. T., van Paassen, H. M. B., Wapenaar, M. C., Grootscholten, P. M., Ginjaar, H. B., Bakker, E., Pearson, P. L., and van Ommen, G. J. B. (1989). High resolution deletion breakpoint mapping in the DMD gene by whole cosmid hybridization. *Nucleic Acids Res.* **17,** 5611–5621.

Bronson, R. T., Lake, B. D., Cook, S., Taylor, S., and Davisson, M. T. (1993). Motor neuron degeneration of mice is a model of neuronal ceroid lipofuscinosis (Batten disease). *Ann. Neurol.* **33,** 381–385.

Bronson, R. T., Donahue, L. R., Johnson, K. R., Tanner, A., Lane, P. W., and Faust, J. R. (1998). Neuronal ceroid lipofuscinosis (*nclf*), a new disorder of the mouse linked to chromosome 9. *Am. J. Med. Genet.* **77,** 289–297.

Callen, D. F., Baker, E., Lane, S., Nancarrow, J., Thompson, A., Whitmore, S. A., MacKennan, D. H., Berger, R., Cherif, D., Jarvela, I., Peltonen, L., Sutherland, G. R., and Gardiner, R. M. (1991). Regional mapping of the Batten disease locus (*CLN3*) to human chromosome 16p12. *Am. J. Hum. Genet.* **49,** 1372–1377.

Church, D. M., Stotler, C. J., Rutter, J. L., Murrell, J. R., Trofatter, J. A., and Buckler, A. J. (1994). Isolation of genes from complex sources of mammalian DNA using exon amplification. *Nat. Genet.* **6,** 98–105.

Cohen, D., Chumakov, I., and Weissenbach, J. (1993). A first-generation physical map of the human genome. *Nature* **266,** 698–701.

Croopnick, J. B., Choi, H. C., and Mueller, D. M. (1998). The subcellular location of the yeast *Saccharomyces cerevisiae* homologue of the protein defective in the juvenile form of Batten disease. *Biochem. Biophys. Res. Commun.* **250,** 335–341.

Dyer, M. R., and Walker, J. E. (1993). Sequences of members of the human gene family for the c subunit of mitochondrial ATP synthase. *Biochem. J.* **293,** 51–64.

Eiberg, H., Gardiner, R. M., and Mohr, J. (1989). Batten disease (Spielmeyer-Sjogren disease) and haptoblobins (HP): Indication of linkage and assignment to chromosome 16. *Clin. Genet.* **36,** 217–218.

Gardiner, R. M., Sandford, A., Deadman, M., Poulton, J., Reeders, S., Jokiaho, I., Peltonen, L., Eiberg, H., and Julier, C. (1990). Batten disease (Spielmeyer-Vogt, juvenile-onset neuronal ceroid lipofuscinosis) gene (*CLN3*) maps to human chromosome 16. *Genomics* **8,** 387–390.

Greene, N. D. E., Bernard, D. L., Taschner, P. E. M., Lake, B. D., de Vos, N., Breuning, M. H., Gardiner, R. M., Mole, S. E., Nussbaum, R. L., and Mitchison, H. M. (1999). A murine model for juvenile NCL: Gene targeting of mouse *Cln3*. *Mol. Genet. Metab.* **66,** 309–313.

Haines, J. L., Boustany, R-M. N., Alroy, J., Auger, K. A., Shook, K. S., Terwedow, H., and Lerner, T. J. (1998). Chromosomal localization of two genes underlying late-infantile neuronal ceroid lipofuscinosis. *Neurogenetics* **1,** 217–222.

International Batten Disease Consortium (1995). Isolation of a novel gene underlying Batten disease, *CLN3*. *Cell* **82,** 949–957.

Jarvela, I. E., Mitchison, H. M., O'Rawe, A. M., Munroe, P. B., Taschner, P. E. M., de Vos, N., Lerner, T. J., D'Arigo, K. L., Callen, D. F., Thompson, A. D., Knight, M., Marrone, B. L., Mundt, M. O., Meincke, L., Breuning, M. H., Gardiner, R. M., Doggett, N. A., and Mole, S. E. (1995). YAC and cosmid contigs spanning the Batten disease (*CLN3*) region of 16p12.1-p11.2. *Genomics* **29,** 478–489.

Katz, M. L., Shibuya, H., Liu, P. C., Kaur, S., Gao, C. L., and Johnson, G. S. (1999). A mouse knockout model for juvenile ceroid-lipofuscinosis (Batten disease). *J. Neurosci. Res.* **57,** 551–556.

Konradi, C., Ozelius, L., Yan, W., Gusella, J. F., and Breakefield, X. O. (1991). Dinucleotide repeat polymorphism on human chromosome 16. *Nucleic Acids Res.* **19,** 5449.

Lee, R. L., Johnson, K. R., and Lerner, T. J. (1996). Isolation and chromosomal mapping of a mouse homolog of the Batten disease gene *CLN3*. *Genomics* **35,** 617–619.

Lerner, T., Wright, G., Dackowski, W., Shook, D., Anderson, M. A., Klinger, K., Callen, D., and Landes, G. (1992). Molecular analysis of human chromosome 16 cosmid clones containing NotI sites. *Mamm. Genome* **3,** 92–100.

Lerner, T. J., Boustany, R-M. N., MacCormack, K., Gleitsman, J., Schlumpf, K., Breakefield, X. O., Gusella, J. F., and Haines, J. L. (1994). Linkage disequilibrium between the juvenile neuronal ceroid lipofuscinosis gene and marker loci on chromosome 16p12.1 *Am. J. Hum. Genet.* **54,** 88–94.

Lerner, T. J., D'Arigo, K. L., Haines, J. L., Doggett, N. A., Taschner, P. E. M., Buckler, A. J., and the Batten Disease Consortium (1995). Isolation of genes from the Batten candidate region using exon amplification. *Am. J. Med. Genet.* **57,** 320–323.

McCormick, M. K., Campbell, E., Deaven, L., and Moyzis, R. (1993). Low-frequency chimeric yeast artificial chromosome libraries from flow-sorted human chromosomes 16 and 21. *Proc. Natl. Acad. Sci. (USA)* **90,** 1063–1067.

Messer, A., Plummer, J., Maskin, P., Coffin, J. M., and Frankel, W. M. (1992). Mapping of the motor neuron degeneration (*Mnd*) gene, a mouse model of amyotrophic lateral sclerosis (ALS). *Genomics* **18,** 7097–802.

Mitchison, H. M., Thompson, A. D., Mully, J. C., Kozman, H. M., Richards, R. I., Callen, D. F., Stallings, R. L., Doggett, N. A., Attwood, J., McKay, T. R., Sutherland, G. R., and Gardiner, R. M. (1993). Fine genetic mapping of the Batten disease locus *CLN3* by haplotype analysis and demonstration of allelic association with chromosome 16p microsatellite loci. *Genomics* **16,** 455–460.

Mitchison, H. M., Taschner, P. E. M., O'Rawe, A. M., De Vos, N., Phillips, H. A., Thompson, A. D., Kozman, H. M., Haines, J. L., Schlumpf, K., D'Arigo, K., Boustany, R-M. N., Callen, D. F., Beuning, M. H., Gardiner, R. M., Mole, S. E., and Lerner, T. J. (1994). Genetic mapping of the Batten disease locus (*CLN3*) to the interval *D16S288-D16S383* by analysis of haplotypes and allelic association. *Genomics* **33,** 465–468.

Mitchison, H. M., O'Rawe, A. M., Taschner, P. E. M., Sandkuijl, L. A., Santavuori, P., de Vos, N., Breuning, M. H., Mole, S. E., Gardiner, R. M., and Jarvela, I. E. (1995). Batten disease gene *CLN3*:

Linkage disequilibrium mapping in the Finnish population and analysis of European haplotypes. *Am. J. Hum. Genet.* **56,** 654–662.

Mitchison, H. M., Munroe, P. B., O'Rawe, A. M., Taschner, P. E. M., de Vos, N., Kremmidiotis, G., Lensink, I., Munk, A. C., D'Arigo, K. L., Anderson, J. W., Lerner, T. J., Moyzis, R. K., Callen, D. F., Breuning, M. H., Doggett, N. A., Gardiner, R. M., and Mole, S. E. (1997). Genomic structure and complete nucleotide sequence of the Batten Disease gene, *CLN3*. *Genomics* **40,** 346–350.

Mitchison, H. M., Bernard, D. J., Greene, N. D., Cooper, J. D., Junaid, M. A., Pullarkat, R. K., de Vos, N., Breuning, M. H., Owens, J. W., Mobley, W. C., Gardiner, R. M., Lake, B. D., Taschner, P. E., and Nussbaum, R. L. (1999). Targeted disruption of the *Cln3* gene provides a mouse model for Batten disease. *Neurobiol. Dis.* **6,** 321–34.

Munroe, P. B., Mitchison, H. M., O'Rawe, A. M., Anderson, J. W., Boustany, R-M. N., Lerner, T. J., Taschner, P. E. M., de Vos, N., Breuning, M. H., Gardiner, R. M., and Mole, S. E. (1997). Spectrum of mutations in the Batten disease gene, *CLN3*. *Am. J. Hum. Genet.* **61,** 310–316.

Nakai, K., and Kanehisa, M. (1992). A knowledge base for predicting protein localization sites in eukaryotic cells. *Genomics* **14,** 897–911.

Palmer, D. N., Fearnley, I. M., Walker, J. E., Hall, N. A., Lake, B. D., Wolfe, L. S., Haltia, M., Martinus, R. D., and Jolly, R. D. (1992). Mitochondrial ATP synthase subunit c storage in ceroid-lipofuscinosis (Batten disease). *Am. J. Med. Genet.* **42,** 561–567.

Pearce, D. A., and Sherman, F. (1997). *BTN1*, a yeast gene corresponding to the human gene responsible for Batten disease, is not essential for viability, mitochondrial function, or degradation of mitochondrial ATP synthase. *Yeast* **13,** 691–697.

Pearce, D. A., and Sherman, F. (1998). A yeast model for the study of Batten disease. *Proc. Natl. Acad. Sci. (USA)* **95,** 6915–6918.

Pearce, D. A., Ferea, T., Nosel, S. A., Das, B., and Sherman, F. (1999). Action of *BTN1*, the yeast orthologue of the gene mutated in Batten disease. *Nat. Genet.* **22,** 55–58.

Ranta, S., Zhang, Y., Ross, B., Lonka, L., Takkunen, E., Messer, A., Sharp, J., Wheeler, R., Kusumi, K., Mole, S., Liu, W., Soares, M. B., Bonaldo, M., Hirvasniemi, A., de la Chapelle, A., Gilliam, T. C., and Lehesjoki, A-E. (1999). The neuronal ceroid lipofuscinosis in the human EPMR and *mnd* mutant mice are associated with mutations in *CLN8*. *Nat. Genet.* **23,** 233–236.

Rogaev, E. I., and Keryanov, S. A. (1992). Unusual variability of the complex dinucleotide repeat block at the *SPN* locus. *Hum. Mol. Genet.* **1,** 657.

Santavuori, P. (1988). Neuronal ceroid lipofuscinosis in childhood. *Brain Dev.* **10,** 80–83.

Savukoski, M., Klockars, T., Homberg, V., Santavuori, P., Lander, E. S., and Peltonen, L. (1998). *CLN5*, a novel gene encoding a putative transmembrane protein mutated in Finnish variant late infantile neuronal ceroid lipofuscinosis. *Nat. Genet.* **19,** 286–288.

Sebald, W., and Hoppe, J. (1981). On the structure and genetics of the proteolipid subunit of the ATP synthase complex. *Curr. Topi. Bioenerget.* **12,** 1–64.

Sharp, J. D., Wheeler, R. B., Lake, B. D., Savukoski, M., Jarvela, I., Peltonen, L., Gardiner, R. M., and Williams, R. E. (1997). Loci for classical and a variant late-infantile neuronal ceroid lipofuscinosis map to chromosome 11p15 and 15q21-23. *Hum. Mol. Genet.* **6,** 591–595.

Shen, Y., Holman, K., Thompson, A., Kozman, H., Callen, D. F., Sutherland, G. R., and Richards, R. I. (1991). Dinucleotide repeat polymorphism at the *D16S288* locus. *Nucleic Acids Res.* **19,** 5445.

Shen, Y., Holman, K., Doggett, N. A., Callen, D. F., Sutherland, G. R., and Richards, R. I. (1993). Four dinucleotide repeat polymorphisms on human chromosome 16. *Hum. Mol. Genet.* **2,** 1745.

Sleat, D. E., Donnelly, R. J., Lackland, H., Leu, C-G., Sohar, I., Pullarkat, R., and Lobel, P. (1997). Association of mutations in a lysosomal protein with classical late-infantile neuronal ceroid lipofuscinosis. *Science* **277,** 1802–1805.

Stallings, R. L., Doggett, N. A., Callen, D., Apostolou, S., Chen, L. Z., Nancarrow, J. K., Whitmore, S. A., Harris, P., Mitchison, H., Breuning, M., Saris, J., Fickett, J., Cinkosky, M., Torney, D. C., Hildebrandt, C. E., and Moyzis, R. K. (1992). Evaluation of a cosmid contig physical map of human chromosome 16. *Genomics* **13,** 1031–1039.

Taschner, P. E. M., de Vos, N., Thompson, A. D., Callen, D. F., Doggett, N. A., Mole, S. E., Dooley, T. P., Barth, P. G., and Beuning, M. H. (1995). Chromosome 16 microdeletion in a patient with juvenile neuronal ceroid lipofuscinosis (Batten disease). *Am. J. Hum. Genet.* **56,** 663–668.

Taschner, P. E. M., de Vos, N., and Breuning, M. H. (1997). Rapid detection of the major deletion in the Batten disease gene *CLN3* by alelle specific PCR. *J. Med. Genet.* **34,** 955–956.

Thompson, A. D., Shen, Y., Holman, K., Sutherland, G. R., Callen, D. F., and Richards, R. I. (1992). Isolation and characterization of $(AC)_n$ microsatellite markers from human chromosome 16. *Genomics* **13,** 402–408.

Vesa, J., Hellsten, E., Verkruyse, L-A., Camp, L. A., Rapola, J., Santavuori, P., Hofmann, S. L., and Peltonen, L. (1995). Mutations in the palmitoyl protein thioesterase gene causing infantile neuronal ceroid lipofuscinosis. *Nature* **376,** 584–587.

Wisniewski, K. E., Rapin, I., and Heaney-Kieras, J. (1988). Clinico-pathological variability in the childhood neuronal ceroid lipofuscinoses and new observations on glycoprotein abnormalities. *Am. J. Med. Genet.* **Suppl. 5,** 27–46.

Yan, W., Boustany, R-M. N., Konradi, C., Ozelius, L., Lerner, T., Troffater, J. A., Julier, C., Breakefield, X. O., Gusella, J. F., and Haines, J. L. (1993). Localization of juvenile, but not late-infantile, neuronal ceroid lipofuscinosis on chromosome 16. *Am. J. Hum. Genet.* **52,** 89–95.

Yan, W. L., Lerner, T. J., Haines, J. L., and Gusella, J. F. (1994). Sequence analysis and mapping of a novel human mitochondrial ATP synthase subunit 9 cDNA (ATP5G3). *Genomics* **24,** 375–377.

Zeman, W. (1974). Studies in the neuronal ceroid lipofuscinosis. *J. Neuropathol. Exp. Neurol.* **33,** 1–12.

6

Studies of Homogenous Populations: CLN5 and CLN8

Susanna Ranta
Department of Molecular Genetics
The Folkhälsan Institute of Genetics
00280 Helsinki, Finland
and
Department of Medical Genetics
University of Helsinki
00014 Helsinki, Finland

Minna Savukoski
Department of Human Molecular Genetics
National Public Health Institute
00300 Helsinki, Finland
and
Department of Medical Genetics
University of Helsinki
00014 Helsinki, Finland

Pirkko Santavuori
Department of Neurology
Hospital for Children and Adolescents
Helsinki University Central Hospital
00029 Helsinki, Finland

Matti Haltia
Department of Pathology
University of Helsinki and Helsinki University Central Hospital
00014 Helsinki, Finland

I. Introduction
II. Clinical Data
   A. Finnish-Variant Late-Infantile NCL
   B. Northern Epilepsy

III. Neurophysiology
      A. Finnish-Variant Late-Infantile NCL
      B. Northern Epilepsy
   IV. Neuroradiology
      A. Finnish-Variant Late-Infantile NCL
      B. Northern Epilepsy
   V. Morphology, Cytochemistry, and Biochemistry
      A. Finnish-Variant Late-Infantile NCL
      B. Northern Epilepsy
   VI. Molecular Genetics and Cell Biology
      A. CLN5
      B. CLN8
   VII. Diagnosis
      A. Finnish-variant Late-Infantile NCL
      B. Northern Epilepsy
   VIII. Treatment
   IX. Mouse Homolog for CLN8
      References

ABSTRACT

Finland and the Finns have been the subject of numerous genetic and genealogical studies, owing to enrichment of certain rare hereditary disorders in the Finnish population. Two types of NCL have so-far been found almost exclusively in Finland: Finnish variant late infantile NCL, vLINCL (CLN5), and the Northern epilepsy syndrome or Progressive epilepsy with mental retardation, EPMR (CLN8). The first symptoms of Finnish vLINCL are concentration problems or motor clumsiness by 3 to 6 years of age, followed by mental retardation, visual failure, ataxia, myoclonus, and epilepsy. Northern epilepsy, the newest member of the NCL family with the most protracted course, is characterized by the onset of generalized seizures between 5 and 10 years of age and subsequent progressive mental retardation. Visual problems are slight and late, while myoclonus has not been observed. Both the Finnish vLINCL and Northern epilepsy are pathologically characterized by intraneuronal cytoplasmic deposits of autofluorescent granules which are Luxol fast blue-, PAS-, and Sudan black B-positive in paraffin sections. In Northern epilepsy the intraneuronal storage process and neuronal destruction are generally of mild degree but highly selective and, in contrast to other forms of childhood onset NCL, the cerebellar cortex is relatively spared. By electron microscopy the storage bodies mainly contain rectilinear complex type and fingerprint profiles in Finnish vLINCL and structures resembling curvilinear profiles in Northern epilepsy. Mitochondrial

ATP synthase subunit c is the main stored protein in both disorders. Both the *CLN5* and *CLN8* genes encode putative membrane proteins with yet unknown functions. Furthermore, a well studied spontaneously occurring autosomal recessive mouse mutant, motor neuron degeneration (mnd) mouse, is a homolog for *CLN8*.

I. INTRODUCTION

The concept of "Finnish disease heritage" refers to a group of more than 30 monogenic diseases, more prevalent in Finland than in other countries (Norio *et al.*, 1973; Peltonen, 1997; de la Chapelle and Wright, 1998). Three neuronal ceroid lipofuscinosis (NCL) subtypes are enriched in Finland and form part of the Finnish disease heritage: infantile NCL (INCL; *CLN1*)(Haltia *et al.*, 1973a, 1973b, Santavuori *et al.*, 1973), Finnish-variant late-infantile NCL (vLINCL; *CLN5*) (Santavuori *et al.*, 1982), and the Northern epilepsy syndrome or progressive epilepsy with mental retardation, EPMR (*CLN8*)(Hirvasniemi *et al.*, 1994, Haltia *et al.*, 1999). The most common NCL subtype in Finland, however, is juvenile NCL (JNCL; *CLN3*) (Santavuori, 1988; Martin *et al.*, 1999). INCL and JNCL are also seen in other parts of the world, whereas, with the exception of one Dutch and one Swedish patient with Finnish vLINCL, Finnish vLINCL and Northern epilepsy have so far been observed only in Finland. Both disorders are rare; to date only 26 patients with Finnish vLINCL and 26 patients with Northern epilepsy have been reported in Finland.

II. CLINICAL DATA

A. Finnish-variant late-infantile NCL

Finnish-variant late-infantile NCL usually presents with visuomotor problems and/or impaired concentration by 3–6 years of age. Fine or gross motor clumsiness and mental decline may begin at the same time or later.

Seizures. In most cases the first epileptic attacks are complex partial seizures debuting between 8 and 9 years of age. Generalized tonic-clonic seizures begin a few months up to 2 years later, while complex partial seizures decrease in frequency. The onset of myoclonus takes place between 7 and 10 years of age. Myoclonic jerks may precipitate seizures or increase after them. Truncal myoclonus interferes with walking ability (Santavuori *et al.*, 1991).

Dementia. Mental deterioration can sometimes be observed already by 3–4 years of age. In most patients, however, mental decline begins some years later. By 6 years of age the intelligence quotient (IQ) may still be normal but the neuropsychological profile shows some deviant features. Most patients with Finnish vLINCL are markedly mentally retarded by 11 years of age.

Neurological findings. At the early stage of the disease, gross and fine motor clumsiness and problems with balance and coordination can be observed. By that time walking ability is still normal. After a couple of years, at the latest by 10 years of age, marked truncal ataxia becomes evident and independent walking ability is soon lost. Dystonic postures may begin at the same stage, speech becomes severely dysarthric, and the first extrapyramidal signs are found.

Ophthalmological findings. In most cases visual failure and macular degeneration are first noticed between 5 and 8 years of age. By that time, electroretinograms also reveal retinal degeneration.

Behavioral problems. Patients with Finnish vLINCL show much less aggressive behavior or irritability than patients with INCL or JNCL (Santavuori et al., 1999). Some patients, however, need medication due to behavioral problems.

Routine laboratory investigations of blood, urine and cerebrospinal fluid are normal. In particular, no vacuolated lymphocytes have been observed. (Santavuori et al., 1982).

B. Northern epilepsy

Seizures. The first presenting symptom in the Northern epilepsy syndrome is epilepsy. Most patients suffer their first seizure between ages 5 and 10, at an average of 6.7 years of age. The first seizures can be triggered by fever. All patients have generalized tonic-clonic seizures, while approximately one-third occasionally also suffer from complex partial seizures during childhood. Almost 40% of patients experience prolonged postictal mental dullness lasting 1–3 days, during which they need help in their daily activities. At the initial stage of the disease, the epileptic attacks occur once a month or 2 months. Toward puberty the frequency of the attacks increases, up to 4–10 a month. During early adulthood the seizures occur less frequently, usually 4–6 times a year. The seizure activity decreases in adulthood, when patients can be seizure-free for several years. However, complete remission has not been described. Northern epilepsy patients do not have myoclonic seizures (Hirvasniemi et al., 1994, 1995).

Dementia. Most patients initially develop normally before the onset of epilepsy. Mental retardation is usually observed 2–5 years after the first seizures. Because of difficulties at school, some patients may need special education classes. Mental decline is most rapid during childhood and puberty, when epileptic activity is most pronounced, but continues slowly throughout life. By the age of 40, all patients are at least moderately retarded (Hirvasniemi and Karumo, 1994; Hirvasniemi et al., 1994, 1995).

Neurological findings. At the onset of the disease, neurological examination does not show any abnormalities. As the disease progresses, difficulties in fine and gross motor tasks are observed. By middle age, patients are clumsy, have difficulties with equilibrium, and their walking is slow and broad-based, with small steps. Most patients have dysphasic speech. One severely affected adult patient completely lost her ability to speak (Hirvasniemi *et al.*, 1995).

Ophthalmological findings. Retinal degeneration, a characteristic finding in NCL, has not been reported in Northern epilepsy. However, some of the patients have diminished (<1.0) visual acuity with no ocular abnormality (Hirvasniemi *et al.*, 1995). Detailed neuroophthalmological studies have not been published.

Behavioral problems. In childhood and puberty nearly half of the patients have behavioral problems, such as irritability, restlessness, disobedience, noisy wakefulness during nights, and inattentiveness. They have often been treated with neuroleptic drugs (Hirvasniemi *et al.*, 1994, 1995).

Routine laboratory investigations of blood, urine, and cerebrospinal fluid show no consistent abnormality. In particular, no vacuolated lymphocytes have been detected. Urinary screenings of amino acids, oligosaccharides, glycosaminoglycans, and organic acids have also given normal results (Hirvasniemi *et al.*, 1994).

III. NEUROPHYSIOLOGY

A. Finnish-variant late-infantile NCL

Giant visual evoked potentials (VEPs), giant somatosensory evoked potentials (SEPs), and posterior spikes to low-frequency photic stimulation in the electroencephalogram (EEG) are characteristic findings in Finnish vLINCL. These findings become visible in an individual order, usually between 7 and 10 years of age, sometimes earlier. Abnormal SEP is often the first finding (Santavuori *et al.*, 1991, 1999).

B. Northern epilepsy

At the onset of Northern epilepsy the EEG recordings are occasionally normal. However, by puberty EEG practically always reveals slowing down of the background activity with a 6 to 7-Hz theta rhythm and impaired reaction to eye opening. At this stage, delta activity is abundant, while sleep-specific potentials can be totally absent. The amount of diffuse delta and theta activities diminishes during the progression of the disease. Clear interictal epileptiform activity is overall relatively scanty but can be observed in nearly half of the EEG recordings.

Nearly half of the patients have abnormal VEPs with delayed latencies. In addition, half of the patients have abnormal nerve conduction velocities (NCV) with no clinical symptoms of neuropathy, and approximately one-third have abnormal brainstem auditory evoked potentials (BAEVs) (Hirvasniemi et al., 1994; Lang et al., 1997).

IV. NEURORADIOLOGY

A. Finnish-variant late-infantile NCL

At an early stage of Finnish vLINCL, magnetic resonance imaging (MRI) reveals high signal rims around the lateral ventricles in the white matter and a lower signal in the thalami than the basal ganglia on T2-weighted images. Moderate cerebellar and slight cerebral atrophy are found. Progressive severe cerebral and particularly cerebellar atrophy are seen during the later course of the disease. The early MRI changes are important for the clinical diagnosis (Autti et al., 1992, 1997).

B. Northern epilepsy

Cerebellar and brainstem atrophy in MRI and computerized tomography (CT) scans are the first abnormal signs appearing in early adulthood. Later, as the disease progresses, cerebral atrophy is also observed. All patients over 30 years of age show evidence of progressive cortical and/or central cerebral atrophy. The degree of the cerebral atrophy correlates with the IQ (Hirvasniemi and Karumo, 1994).

V. MORPHOLOGY, CYTOCHEMISTRY, AND BIOCHEMISTRY

A. Finnish-variant late-infantile NCL

Macroscopic autopsy findings. All autopsied patients have shown severe generalized cerebral and extreme cerebellar atrophy (Figure 6.1.), with brain weights of the order of 500–700 g. The cortical ribbon is thin and the white matter is reduced in amount, grayish, and tough. There is pronounced ventricular dilatation. No macroscopic extraneural lesions have been observed (Tyynelä et al., 1997).

Histology. The most characteristic feature, observed at all ages studied, is the presence of storage granules in the cytoplasm of most nerve cells. Small amounts of similar storage material may be seen in many other cell types throughout the body. Biopsy and autopsy studies of patients at different ages suggest that

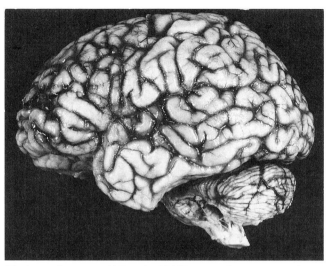

Figure 6.1. The brain of a 14-year-old Finnish vLINCL patient shows pronounced generalized atrophy, particularly striking in the cerebellum (brain weight 655 g).

the most pronounced storage is initially found in the large pyramidal cells of the deeper cortical layers. At this stage, the neurons of the superficial part of lamina III characteristically show frequent meganeurites filled with storage granules (Figure 6.2A). During later stages, there is progressive loss of neurons, particularly in laminae III and V, with occasional figures of neuronophagy and pronounced astrocytic hyperplasia and hypertrophy (Tyynelä et al., 1997).

In the hippocampus the sectors CA2–CA4 show pronounced intraneuronal storage, while CA1 is only mildly affected. In the cerebellar cortex there is an almost total destruction of the granule cells, while an occasional preserved Purkinje cell may be seen with dilated dendrites loaded with storage granules. There is pronounced proliferation and hypertrophy of the Bergmann astrocytes.

The subcortical structures show moderate to pronounced intraneuronal storage but, with the exception of the thalami, only mild to moderate neuronal loss. There is advanced loss of myelin in the cerebral and cerebellar white matter, accompanied by astrocytic proliferation and hypertrophy (Tyynelä et al., 1997).

Ultrastructure. At electron microscopy the storage granules correspond to electron-dense cytosomes surrounded by a lysosomal membrane and with a variable internal ultrastructure. In brain biopsies and autopsies some intraneuronal cytosomes show features resembling fingerprint or curvilinear profiles, as seen in patients with classic LINCL (cLINCL) or JNCL, while the ultrastructure of most cytosomes corresponds to the so-called rectilinear complex (Figure 6.2B)

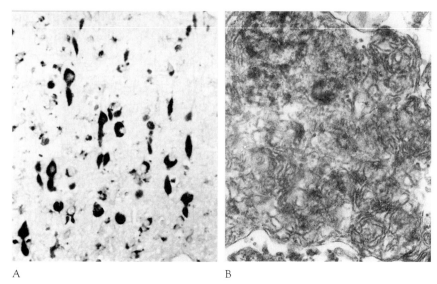

Figure 6.2. (A) Cerebral cortical nerve cells show storage granules immunoreactive for the subunit c of the mitochondrial ATP synthase. Note the ballooning of the cell bodies and the meganeurites filled with the storage material. Paraffin section, immunoperoxide stain for subunit c of the mitochondrial ATP synthase ($\times 300$). (B) The intraneuronal membrane-bound storage bodies in Finnish vLINCL consist largely of membraneous profiles with an ultrastructure corresponding to the rectilinear complex (electron micrograph $\times 20,000$).

(Santavuori et al., 1982; Tyynelä et al., 1997; Elleder et al., 1999). In skin and rectal mucosal biopsies, non-neuronal cells (endothelial, smooth muscle, and Schwann cells) predominantly show inclusions of atypical curvilinear or rectilinear complex type, while autonomic ganglion cells of the vermiform appendix and rectum mainly contain fingerprint bodies (Santavuori et al., 1982). Blood lymphocytes contain cytosomes with variegated structures including fingerprint profiles. These inclusions are almost identical to those seen in patients with other vLINCLs (CLN6 and CLN7) (Rapola and Lake, 2000).

Cytochemistry and biochemistry. The storage granules are relatively resistant to lipid solvents and stain positively with the Luxol fast blue, PAS and Sudan black B stains in paraffin sections. They are autofluorescent when viewed in ultraviolet light. They bind antibodies raised against the subunit c of the mitochondrial ATP synthase (Figure 6.2A) and the sphingolipid activator proteins A and D. However, isolation and purification of the cerebral storage granules and subsequent quantitative sequence analysis showed subunit c to be the major component of the pathological cytosomes, with only minor accumulation of the sphingolipid activator proteins (Tyynelä et al., 1997).

B. Northern epilepsy

Macroscopic autopsy findings. The only disease-related macroscopic abnormality noted in three autopsied patients was moderate generalized cerebral gyral atrophy in one of them (Herva et al., 2000).

Histology. All three autopsied patients showed deposits of granular storage material in the cytoplasm of most neurons and, to a much lesser degree, in many other cell types throughout the body, including, e.g., heart muscle cells, Kupffer cells of the liver, and distal convoluted tubules of the kidney (Herva et al., 2000). Although the intraneuronal storage was almost ubiquitous, the amount of the stored material varied greatly, resulting in a characteristic, highly selective distributional pattern. In the cerebral neocortex the large pyramidal neurons of lamina III showed pronounced ballooning of their cell bodies, while the less ballooned neurons of the upper parts of this lamina displayed numerous axonal spindles (meganeurites), distended by the storage granules. Other neocortical laminae were much less affected, with mild patchy neuronal loss in lamina V. In the hippocampus the sectors CA2 (Figure 6.3A), CA3, and CA4 showed pronounced storage, with some neuronal loss in CA2, while CA1 was almost intact (Figure 6.3B). Slight to moderate intraneuronal storage occurred in the basal ganglia and thalamus, most brainstem nuclei, as well as in the spinal cord. However, the pigmented brainstem nuclei were almost intact. In sharp contrast to other childhood-onset forms of NCL, even the cerebellar cortex was relatively preserved. Occasional figures of neuronophagy and some astrocytic hyperplasia and hypertrophy were seen in the hippocampal CA2 field, and modest patchy astrocytic proliferation even in the deeper neocortical layers. The white matter was relatively spared. Studies of blood smears did not disclose any vacuolated lymphocytes (Haltia et al., 1999; Herva et al., 2000).

Ultrastructure. The membrane-bound electron-dense storage cytosomes contain loosely packed curved lamellar structures, resembling curvilinear profiles

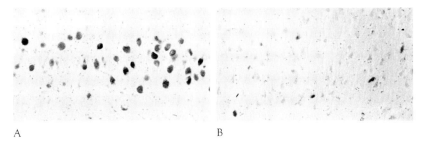

A   B

Figure 6.3. In Northern epilepsy patients the hippocampal sector CA2 shows pronounced intraneuronal storage (A), while the CA1 field is only minimally affected (B). Paraffin section, immunoperoxidase stain for subunit c of the mitochondrial ATP synthase (×200).

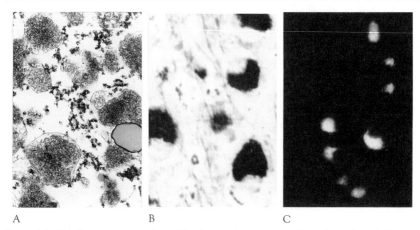

Figure 6.4. (A) The storage granules in Northern epilepsy consist of membrane-bound electron-dense aggregates with an internal ultrastructure resembling curvilinear profiles (electron micrograph ×10,000). In paraffin sections the intraneuronal storage bodies are Sudan black B-positive (B) and show strong autofluorescence in ultraviolet light (C).

(Figure 6.4A). In addition, some electron-dense granular material may be seen (Herva et al., 2000).

Cytochemistry and biochemistry. The intracytoplasmic storage granules are Luxol fast blue-, PAS-, and Sudan black B-positive in paraffin sections (Figure 6.4B), and they are autofluorescent in ultraviolet light (Figure 6.4C). They are immunoreactive for subunit c of the mitochondrial ATP synthase (Figure 6.3A) and sphingolipid activator proteins, and also show amyloid-beta-like immunoreactivity. However, N-terminal sequence analysis of purified storage material reveals only one major sequence, indicating that subunit c is the major storage protein (Herva et al., 2000).

VI. MOLECULAR GENETICS AND CELL BIOLOGY

A. CLN5

Isolation of the CLN5 gene. The CLN5 gene was assigned to chromosome 13q21-32 (Savukoski et al., 1994). Subsequently, the region was refined by haplotype and linkage disequilibrium analyses to 200 kb and a physical map was constructed across the disease gene region (Klockars et al., 1996). The CLN5 gene was initially identified by traditional positional cloning using a fetal brain cDNA library screening. The assembly of the CLN5 sequence was expedited by large-scale sequencing of the critical region (Savukoski et al., 1998).

Table 6.1. Mutations in CLN5 and CLN8

Nucleotide change	Mutation type	Amino acid change/predicted consequence	Location	Nationality of the patients	Reference
CLN5					
1517G > A	Nonsense	W75X	Exon 1	Finnish	Savukoski et al., 1998
2467delAT	2-bp Deletion	Y392X	Exon 4	Finnish	Savukoski et al., 1998
2127G > A	Missense	D279N	Exon 4	Dutch	Savukoski et al., 1998
CLN8					
C.70C > G	Missense	R24G	Exon 2	Finnish	Ranta et al., 1999
Cln8 (Mnd mouse)					
267-268insC	1-bp Insertion	Frameshift after P89, Truncated 116 aa protein			Ranta et al., 1999

Gene structure. The 4.1-kb *CLN5* gene consists of four exons and contains a coding region of 1221 bp, predicting a polypeptide of 407 amino acids. In the genomic DNA, *CLN5* spans approximately 13 kb (Savukoski et al., 1998).

Mutations in CLN5. So far, three mutations have been identified in *CLN5* (See Table 6.1). The most frequently seen mutation, *CLN5* Fin major, was observed in 94% of the Finnish disease chromosomes. It is a 2-bp deletion (2467delAT or C.1175delAT) in exon 4, resulting in a substitution of tyrosine to STOP at codon 392 (Y392X) and leading to a predicted truncated protein of 391 amino acids. The *CLN5* Fin major mutation has also been observed in heterozygous form in a single Finnish patient with a fourth, yet-unidentified mutation (Rapola et al., 1999). The second known mutation, *CLN5* Fin minor, has been observed homozygous in a Finnish family with two affected siblings. In addition, a Swedish patient with a single *CLN5* Fin minor mutation and an unidentified mutation in the second allele has been documented (Savukoski, 1999). A 1-bp substitution (1517G > A or C.225G > A) results in a nonsense mutation (W75X) and a predicted polypeptide of only 74 amino acids. The third and only known mutation of non-Finnish origin, *CLN5* European, a missense change (2127G > A or C.835G > A) resulting in an aspartic acid-to-asparagine substitution at codon 279 (D279N), was detected homozygous in a single Dutch patient. Screening of 700 controls originating from the high-risk region on the west coast of Finland for the *CLN5* Fin major mutation revealed a local carrier frequency of 1:24–1:100 (Savukoski et al., 1998). The carrier frequency of the *CLN5* Fin major mutation varies between 1:300 and 1:1000 elsewhere in Finland (Tomi Pastinen, personal communication).

Tissue expression of the CLN5 gene. Transcripts of 2.0, 3.0, and 4.5 kb were detected in all tissues tested when the tissue expression pattern of CLN5 was monitored by Northern hybridization. An additional 5.5-kb signal was seen in skeletal muscle. Of the transcripts identified, the 4.5 signal corresponds best to the assembled 4.1 kb CLN5 cDNA. The other signals may represent alternatively spliced transcripts. Among adult tissues, the level of expression was highest in aorta, kidney, lung, and pancreas. A fairly uniform expression was seen in all fetal tissues (brain, heart, kidney, liver, spleen, lung), except for fetal thymus, which showed approximately two times higher signal intensity (Savukoski et al., 1998).

Computer predictions. No previously known genes or proteins homologous to CLN5 were identified through database searches. Several potential modification sites, including protein kinase C phosphorylation sites, casein kinase II phosphorylation sites, tyrosine kinase phosphorylation sites, N-myristoylation sites, and N-glycosylation sites were identified. The molecular weight predicted from the 407-amino acid sequence of CLN5 is 46 kD, and the calculated pI 8.41. Based on hydrophobicity predictions, CLN5 was suggested to be a transmembrane protein with two transmembrane domains and an intraluminar loop. The location of the CLN5 protein within the cell is unknown (Savukoski et al., 1998).

B. CLN8

Isolation of the CLN8 gene. The CLN8 gene was assigned to the telomeric region of chromosome 8p (Tahvanainen et al., 1994). A yeast artificial chromosome (YAC) contig was then constructed across the 4-cM genetic region harboring the disease locus (Ranta et al., 1996). The critical gene region was further refined to less than 700 kb and a bacterial artificial chromosome (BAC) contig across the critical region was completed (Ranta et al., 1997). The CLN8 gene was identified by positional cloning using a YAC insert to screen a normalized brain cDNA library (Ranta et al., 1999).

Gene structure. The total length of the assembled cDNA sequence is 4819 bp, including a coding region of 861 bp, predicting a protein of 286 amino acids. The CLN8 gene has three exons. The 5' untranslated region consists of a noncoding exon 1 and 123 bp of the 5' sequence of exon 2. The coding region is divided into exons 2 and 3. The total length of exon 3 as well as the intron sizes are unknown. The 3' untranslated region contains at least four poly-A tails suggesting alternative splicing (Ranta et al., 1999).

The CLN8 mutation. A C.70C > G change in exon 2 resulting in an arginine-to-glycine substitution at codon 24 (R24G) was found to fully co-segregate with the disease phenotype (Table 6.1). Screening of 271 unrelated Finnish controls for the C.70C > G mutation revealed two carriers, giving a carrier frequency of 1:135 in Finland. Additional screening of 92 controls from the high-risk Kainuu region revealed a local carrier frequency of 1:46, strongly suggesting that this is the gene defect underlying Northern epilepsy (Ranta et al., 1999).

Tissue expression of the CLN8 gene. When the tissue expression pattern of CLN8 was studied by Northern hybridizations, RNA transcripts of 1.4, 3.4, and 7.5 kb were detected in all tissues. Of the transcripts identified by Northern analysis, the 1.4-kb signal corresponds most closely to the first poly-A tail at 1.3 kb and the 3.4 kb signal to the third (at 3.23 kb) and fourth (at 3.28 kb) poly-A tails. The other signals most likely represent alternatively spliced transcripts. Expression was also detected in fetal tissues (Ranta et al., 1999).

Computer predictions. Although several identical human and mouse ESTs were identified, database searches revealed no previously known genes or proteins homologous to CLN8 (Ranta et al., 1999). Several potential modification sites, such as an N-glycosylation site, two protein kinase C phosphorylation sites, two casein kinase II phosphorylation sites, five N-myristoylation sites, a prokaryotic membrane lipoprotein lipid attachment site, and a leucine zipper pattern were identified. The molecular weight predicted from the polypeptide sequence is 33 kD and the predicted p*I* is 9.6. Based on hydrophobicity predictions, CLN8 was suggested to be a transmembrane protein with several transmembrane domains.

Intracellular localization. Evidence from cell biological studies strongly suggest that CLN8 is an endoplasmic reticulum (ER) resident protein which recycles between ER and ER-Golgi intermediate compartment (Lonka et al., 2000).

VII. DIAGNOSIS

The diagnosis of NCL has traditionally been based on careful clinical evaluation of the patient, neurophysiological studies, and demonstration of characteristic inclusions in tissue specimens observed by light and electron microscopy. Today, DNA-based genetic tests or assays for specific enzyme activities are also available for some of the NCL types.

A. Finnish-variant late-infantile NCL

Finnish vLINCL (CLN5) is clinically a variant of the late-infantile NCL (LINCL) with onset usually between ages 3 and 6 (Santavuori et al., 1982, 1993). Although the age of onset can overlap with the classic LINCL (cLINCL; CLN2), Finnish vLINCL can be distinguished from cLINCL, based on milder clinical course, differences in ultrastructure of the pathological inclusions (Santavuori et al., 1982) as well as genetic analyses or enzyme assay for CLN2. The phenotypes of Finnish

Table 6.2. Characteristics of Finnish vLINCL and Northern Epilepsy

	Finnish vLINCL	Northern epilepsy
Age at onset	3–6 years	5–10 years
Presenting symptom	Visuomotor and concentration problems; motor clumsiness and muscular hypotonia	Seizures
Seizure type	Generalized tonic-clonic (all patients) Partial (may occur at onset)	Generalized tonic-clonic (all patients) Partial (30% of the patients during the early years)
Mental retardation	Present, progressive	Present, progressive
Visual failure	Early, progressive	Not a prominent feature
EEG	Progressive slowing of background activity, low-rate intermittent photic spikes	Progressive slowing of background activity, scanty interictal epileptiform activity
Myoclonus	Present	Not present
Life expectancy	Around 20 years of age (13–35 years)	Late middle age
Electron microscopy	Mainly rectilinear complex and fingerprint profiles	Structures resembling curvilinear profiles and granular material

vLINCL (see Table 6.2) and the other two variants of late-infantile NCL (CLN6, CLN7) can overlap. However, the clinical course in CLN6 is usually more rapid than in Finnish vLINCL. The clinical phenotype together with typical MRI findings, lack of vacuolated lymphocytes, and the ophthalmological findings strongly support diagnosis of Finnish vLINCL. The nationality of the patient should also be taken into account while considering differential diagnosis between the variant LINCLs. Furthermore, as the gene underlying Finnish vLINCL is now known, the diagnosis can usually be confirmed by DNA analysis.

B. Northern epilepsy

Children with Northern epilepsy are usually first examined by a pediatric neurologist because of epilepsy which may be difficult to control with conventionally used anticonvulsants. A diagnosis other than idiopathic epilepsy is often suspected only after mental decline has become evident. If a child with intractable epilepsy has no known etiology for the seizures, no definite MRI findings, and within a few years shows mental decline, the possibility of Northern epilepsy should be considered. At our present stage of knowledge the only possibility for confirming the Northern epilepsy diagnosis relies on DNA analyses. The age of onset of the Northern epilepsy syndrome overlaps with that of JNCL. However, differential diagnosis between these two NCL types is easy: contrary to Northern epilepsy, vacuolated lymphocytes and visual failure combined with macular and retinal degeneration are characteristic findings in JNCL, and epilepsy is seldom the first sign of disease. The clinical picture and course of the disease as well as MRI findings distinguish Northern epilepsy, as seen in the Finnish patients (see Table 6.2), from all variant LINCLs.

VIII. TREATMENT

The treatment for Finnish vLINCL and Northern epilepsy is symptomatic, and there is no known cure. Parents and other members of the family need guidance and support in dealing with everyday aspects of these rare but devastating neurodegenerative disorders. Bone marrow transplantations are not expected to be beneficial with Finnish vLINCL or Northern epilepsy and have not been attempted. In addition to antiepileptic drugs, some patients need medication for behavioral and neurological problems during childhood and puberty.

Anticonvulsants. At the early stage of Finnish vLINCL, monotherapy with lamotrigine or valproate is recommended. Later on, these drugs are often used in combination. Some patients may need benzodiazepine (clobazam or clonazepam) as a third antiepileptic drug. Most patients with Finnish vLINCL also require baclophen and tizanidine to prevent myoclonic jerks and dystonia and to reduce the development of contractures.

In Northern epilepsy, clonazepam has had the best antiepileptic effect. In puberty, when seizures are most frequent, clonazepam has produced seizure-free periods of up to 3 years. The benefit of clonazepam is also evident in the EEG recordings, especially in childhood and puberty, but also in middle age. Only a transient response was achieved with valproate and phenobarbital, whereas phenytoin and carbamazepine proved ineffective. So far there is no experience of the use of lamotrigine on Northern epilepsy (Hirvasniemi *et al.*, 1994, 1995).

IX. MOUSE HOMOLOG FOR CLN8

A spontaneously occurring well-characterized autosomal recessive NCL mutant, the motor neuron degeneration (*mnd*) mouse, was originally considered a model for amyotrophic lateral sclerosis (Messer and Flaherty, 1986). Previous histopathological studies on *mnd* had revealed accumulation of autofluorescent inclusions immunoreactive with antibodies to the subunit c of the mitochondrial ATP synthase indicating that *mnd* is a model for NCL (Bronson *et al.*, 1993; Messer and Plummer, 1993; Pardo *et al.*, 1994). Mnd mice experience early onset retinal and adult-onset progressive motor system degeneration (Messer and Flaherty, 1986;

Messer et al., 1993). The motor dysfunction leads to spastic limb paralysis and premature death. The severity of the phenotype depends also on the genetic background of the mice (Messer et al., 1995).

The sequence of the mouse gene (Cln8) corresponding to the human CLN8 was assembled from ESTs homologous to the human gene. At the nucleotide level, the coding sequences of the human and mouse genes were found to be 82% identical. At the polypeptide level an 85% identity was revealed. The CLN8 mutation site was found to be conserved in the mouse. The mouse Cln8 gene was assigned to the same region on mouse chromosome 8 as mnd. A homozygous 1-bp insertion (267-268insC) in the Cln8 gene resulting in a frameshift and a predicted stop codon 27 codons downstream was identified using mnd mouse DNA. This finding verifies that the previously identified human gene underlies Northern epilepsy and that mnd is a naturally occurring animal model for this disorder (Ranta et al., 1999).

Interestingly, unlike patients with Northern epilepsy, mnd mice do not suffer from epileptic seizures. The motor dysfuction and visual failure, severe in mnd, are also not prominent features in the Northern epilepsy syndrome. The differences between the clinical presentations may partly be explained by the difference between the gene defects, a missense mutation leading to Northern epilepsy syndrome and a frame shift and a truncated protein leading to mnd. Furthermore, more severe mutations in the human gene might result in a strikingly different clinical phenotype.

References

Autti, T., Raininko, R., Launes, J., Nuutila, A., and Santavuori, P. (1992). Jansky-Bielschowsky variant disease: CT, MRI, and SPECT findings. *Pediatr. Neurol.* **8,** 121–126.
Autti, T., Raininko, R., Vanhanen, S. L., and Santavuori, P. (1997). Magnetic resonance techniques in neuronal ceroid lipofuscinoses and some other lysosomal diseases affecting the brain. *Curr. Opin. Neurol.* **10,** 519–524.
Bronson, R. T., Lake, B. D., Cook, S., Taylor, S., and Davisson, M. T. (1993). Motor neuron degeneration of mice is a model of neuronal ceroid lipofuscinosis (Batten's disease). *Ann. Neurol.* **33,** 381–385.
de la Chapelle, A., and Wright, F. A. (1998). Linkage disequilibrium mapping in isolated populations: The example of Finland revisited. *Proc. Natl. Acad. Sci. (USA)* **95,** 12416–12423.
Elleder, M., Lake, B. D., Goebel, H. H., Rapola, J., Haltia, M., and Carpenter, S. (1999). Definitions of the Ultrastructural Patterns found in NCL. In: "The Neuronal Ceroid Lipofuscinoses (Batten Disease)" (H. H. Goebel, S. E. Mole, and B. D. Lake, eds.), pp. 5–15. IOS Press, Amsterdam, The Netherlands.
Haltia, M., Rapola, J., and Santavuori, P. (1973a). Infantile type of so-called neuronal ceroid-lipofuscinosis. Histological and electron-microscopic studies. *Acta Neuropathol.* **26,** 157–170.
Haltia, M., Rapola, J., Santavuori, P., and Keränen, A. (1973b). Infantile type of so-called neuronal ceroid-lipofuscinosis. Morphological and biochemical studies. *J. Neurol. Sci.* **18,** 269–285.
Haltia, M., Tyynelä, J., Hirvasniemi, A., Herva, R., Ranta, U. S., and Lehesjoki, A.-E. (1999). CLN8:

Northern epilepsy. In: "The Neuronal Ceroid Lipofuscinoses (Batten's Disease)" (H. H. Goebel, S. E. Mole, and B. D. Lake, eds.), pp. 117–121. IOS Press, Amsterdam, The Netherlands.

Herva, R., Tyynelä, J., Hirvasniemi, A., Syrjäkallio-Ylitalo, M., and Haltia, M. (2000). Northern epilepsy: A novel form of neuronal ceroid-lipofuscinosis. *Brain Pathol.* **10,** 215–222.

Hirvasniemi, A., and Karumo, J. (1994). Neuroradiological findings in the northern epilepsy syndrome. *Acta Neurol. Scand.* **90,** 388–393.

Hirvasniemi, A., Lang, H., Lehesjoki, A. E., and Leisti, J. (1994). Northern epilepsy syndrome: An inherited childhood onset epilepsy with associated mental deterioration. *J. Med. Genet.* **31,** 177–182.

Hirvasniemi, A., Herrala, P., and Leisti, J. (1995). Northern epilepsy syndrome: Clinical course and the effect of medication on seizures. *Epilepsia.* **36,** 792–797.

Klockars, T., Savukoski, M., Isosomppi, J., Laan, M., Järvelä, I., Petrukhin, K., Palotie, A., and Peltonen, L. (1996). Efficient construction of a physical map by fiber-FISH of the CLN5 region: Refined assignment and long-range contig covering the critical region on 13q22. *Genomics* **35,** 71–78.

Lang, A. H., Hirvasniemi, A., and Siivola, J. (1997). Neurophysiological findings in the northern epilepsy syndrome. *Acta Neurol. Scand.* **95,** 1–8.

Lonka, L., Kyttälä, A., Ranta, S., Jalanko, A., and Lehesjoki, A.-E. (2000). The neuronal ceroid lipofuscinosis CLN8 membrane protein is a resident of the endoplasmic reticulum. *Hum. Mol. Genet.* **9,** 1691–1697.

Martin, J.-J., Ceuterick, C., Elleder, M., Kraus, J., Nevsimalová, S., Sixtová, K., Zeman, J., Santavuori, P., Chabrol, B., Kohlschütter, A., Lake, B. D., Christomanou, H., Cardona, F., Nardocci, N., Kmiec, T., Pineda, M., Milá, M., Ferrer, I., Navarro, C., Uvebrant, P., Hofman, I., Taschner, P. E. M., Williams, R. E., Topçu, M., and Çaliskan, M. (1999). NCL in different European countries. In: "The Neuronal Ceroid Lipofuscinoses (Batten Disease)" (H. H. Goebel, S. E. Mole, and B. D. Lake, eds.), pp. 128–141. IOS Press, Amsterdam, The Netherlands.

Messer, A., and Flaherty, L. (1986). Autosomal dominance in a late-onset motor neuron disease in the mouse. *J. Neurogenet.* **3,** 345–355.

Messer, A., and Plummer, J. (1993). Accumulating autofluorescent material as a marker for early changes in the spinal cord of the Mnd mouse. *Neuromuscul. Disord.* **3,** 129–134.

Messer, A., Plummer, J., Wong, V., and Lavail, M. M. (1993). Retinal degeneration in motor neuron degeneration (mnd) mutant mice. *Exp. Eye Res.* **57,** 637–641.

Messer, A., Plummer, J., MacMillen, M. C., and Frankel, W. N. (1995). Genetics of primary and timing effects in the mnd mouse. *Am. J. Med. Genet.* **57,** 361–364.

Norio, R., Nevanlinna, H. R., and Perheentupa, J. (1973). Hereditary diseases in Finland; rare flora in rare soul. *Ann. Clin. Res.* **5,** 109–141.

Pardo, C. A., Rabin, B. A., Palmer, D. N., and Price, D. L. (1994). Accumulation of the adenosine triphosphate synthase subunit C in the mnd mutant mouse. A model for neuronal ceroid lipofuscinosis. *Am. J. Pathol.* **144,** 829–835.

Peltonen, L. (1997). Molecular background of the Finnish disease heritage. *Ann. Med.* **29,** 553–556.

Ranta, S., Lehesjoki, A. E., Hirvasniemi, A., Weissenbach, J., Ross, B., Leal, S. M., de la Chapelle, A., and Gilliam, T. C. (1996). Genetic and physical mapping of the progressive epilepsy with mental retardation (EPMR) locus on chromosome 8p. *Genome Res.* **6,** 351–360.

Ranta, S., Lehesjoki, A. E., de Fatima Bonaldo, M., Knowles, J. A., Hirvasniemi, A., Ross, B., de Jong, P. J., Soares, M. B., de la Chapelle, A., and Gilliam, T. C. (1997). High-resolution mapping and transcript identification at the progressive epilepsy with mental retardation locus on chromosome 8p. *Genome Res.* **7,** 887–896.

Ranta, S., Zhang, Y., Ross, B., Lonka, L., Takkunen, E., Messer, A., Sharp, J., Wheeler, R., Kusumi, K., Mole, S., Liu, W., Soares, M. B., Bonaldo, M. F., Hirvasniemi, A., de la Chapelle, A., Gilliam,

T. C., and Lehesjoki, A. E. (1999). The neuronal ceroid lipofuscinoses in human EPMR and mnd mutant mice are associated with mutations in CLN8. *Nat. Genet.* **23,** 233–236.

Rapola, J., and Lake, B. (2000). Lymphocyte inclusions in Finnish variant late infantile neuronal ceroid lipofuscinosis (CLN5). *Neuropediatrics* **31,** 33–34.

Rapola, J., Lähdetie, J., Isosomppi, J., Helminen, P., Penttinen, M., and Järvelä, I. (1999). Prenatal diagnosis of variant late infantile neuronal ceroid lipofuscinosis (vLINCL; CLN5). *Prenat. Diagn.* **19,** 685–688.

Santavuori, P. (1988). Neuronal ceroid-lipofuscinoses in childhood. *Brain. Dev.* **10,** 80–83.

Santavuori, P., Haltia, M., Rapola, J., and Raitta, C. (1973). Infantile type of scalled neuronal ceroid-lipofuscinosis. A clinical study of 15 patients. *J. Neurol. Sci.* **18,** 257–267.

Santavuori, P., Rapola, J., Sainio, K., and Raitta, C. (1982). A variant of Jansky-Bielschowsky disease. *Neuropediatrics* **13,** 135–141.

Santavuori, P., Rapola, J., Nuutila, A., Raininko, R., Lappi, M., Launes, J., Herva, R., and Sainio, K. (1991). The spectrum of Jansky-Bielschowsky disease. *Neuropediatrics* **22,** 92–96.

Santavuori, P., Rapola, J., Raininko, R., Autti, T., Lappi, M., Nuutila, A., Launes, J., and Sainio, K. (1993). Early juvenile neuronal ceroid-lipofuscinosis or variant Jansky-Bielschowsky disease: Diagnostic criteria and nomenclature. *J. Inherit. Metab. Dis.* **16,** 230–232.

Santavuori, P., Rapola, J., Haltia, M., Tyynelä, J., Peltonen, L., and Mole, S. E. (1999). CLN5 Finnish Variant Late Infantile NCL. In: "The Neuronal Ceroid Lipofuscinoses (Batten Disease)" (H. H. Goebel, S. E. Mole, and B. D. Lake, eds.), pp. 91–101. IOS Press, Amsterdam, The Netherlands.

Savukoski, M. (1999). Molecular genetics of the late infantile neuronal ceroid lipofuscinosis (LINCL): One gene (CLN5) and two gene loci (CLN2 and CLN6). Academic dissertation, National Public Health Institute.

Savukoski, M., Kestilä, M., Williams, R., Järvelä, I., Sharp, J., Harris, J., Santavuori, P., Gardiner, M., and Peltonen, L. (1994). Defined chromosomal assignment of CLN5 demonstrates that at least four genetic loci are involved in the pathogenesis of human ceroid lipofuscinoses. *Am. J. Hum. Genet.* **55,** 695–701.

Savukoski, M., Klockars, T., Holmberg, V., Santavuori, P., Lander, E. S., and Peltonen, L. (1998). CLN5, a novel gene encoding a putative transmembrane protein mutated in Finnish variant late infantile neuronal ceroid lipofuscinosis. *Nat. Genet.* **19,** 286–288.

Tahvanainen, E., Ranta, S., Hirvasniemi, A., Karila, E., Leisti, J., Sistonen, P., Weissenbach, J., Lehesjoki, A. E., and de la Chapelle, A. (1994). The gene for a recessively inherited human childhood progressive epilepsy with mental retardation maps to the distal short arm of chromosome 8. *Proc. Natl. Acad. Sci. (USA)* **91,** 7267–7270.

Tyynelä, J., Suopanki, J., Santavuori, P., Baumann, M., and Haltia, M. (1997). Variant late infantile neuronal ceroid-lipofuscinosis: Pathology and biochemistry. *J. Neuropathol. Exp. Neurol.* **56,** 369–375.

7

Molecular Genetic Testing for Neuronal Ceroid Lipofuscinoses

Nanbert Zhong[*]
Molecular Neurogenetic Diagnostic Laboratory
New York State Institute for Basic Research in Developmental Disabilities
Staten Island, New York 10314
and
Department of Neurology
SUNY Health Science Center at Brooklyn
Brooklyn, New York 11203

    I. Introduction
   II. Specimens Required for Genetic Testing
      A. Peripheral Blood
      B. Tissues
  III. Molecular Genetic Testing for JNCL
      A. Molecular Testing for the 1.02-kb Common Deletion in the CLN_3 Gene
      B. Molecular Detection of Uncommon Mutations
  IV. Molecular Genetic Testing for LINCL and INCL
      A. Molecular Testing for Common Mutations in INCL and LINCL
      B. Gene Scan of Uncommon Mutations in LINCL
   V. Molecular Screening of Carrier Status in NCL Families
      A. Family Members of Individuals with NCL are Encouraged to be Screened for Carrier Status
      B. Characterize Familial Mutations before Screening for Carrier Status is Requested
  VI. Prenatal Diagnostic Testing for NCL
      A. Specimens Used for Prenatal Molecular Testing

[*]Address for correspondence: E-mail: omrddzhong@aol.com

B. Information about Familial Mutation(s) is a Basic
      Requirement for Prenatal Molecular Genetic Testing
   C. Molecular Approaches for Prenatal Diagnosis of NCL
VII. Important Issues in the Molecular Genetic Testing
   A. Testing Five Common Mutations in CLN_1, CLN_2,
      and CLN_3 may Detect 70% of Classical INCL, LINCL,
      and JNCL Cases
   B. Clinical and Pathological Data Should be Collected
      before Genetic Testing is Requested
   C. Molecular Screening of Carrier Status in NCL-Affected
      Families is Encouraged
   References

ABSTRACT

Eight different NCL forms have been recognized to be encoded by genes CLN_{1-8}. $CLN_{1,2,3,5, and 8}$ have been cloned, and at least 85 mutations have been detected. Molecular technology can now be applied to genetic testing for NCLs; testing is now available in clinic diagnostic and research laboratories for *CLN* genes that have been cloned. Molecular genetic testing makes it possible not only to confirm clinical and pathological diagnoses but also to offer pre-symptom diagnosis and carrier screening for NCL families. In addition, DNA-based mutation analysis may predict prenatal outcome more accurately for pregnant women in NCL families.

I. INTRODUCTION

The neuronal ceroid lipofuscinoses (NCLs) are a group of newly categorized lysosomal storage disorders with autosomal recessive inheritance. Eight different NCL forms have been recognized in this group, which result from not only the deficiency of lysosomal enzymes but also the dysfunction of structural proteins (Mole, 1998; Bennett and Hofmann, 1999; Zhong, 2000). Clinically, NCLs are characterized by progressive deterioration of vision, seizures, cognitive and motor dysfunction, and involuntary movement and developmental disabilities. Three classical NCL forms, the infantile (INCL), late-infantile (LINCL), and juvenile (JNCL), are among the most common childhood-onset neurodegenerative disorders, with an estimated incidence of 1/12,500 to 1/100,000 globally (Mole, 1998). In the United States, more than 440,000 families are affected by NCL (Rider and Rider, 1997), and this figure has increased in recent years with the progress of molecular genetic testing.

Clinically, the first presenting symptom in NCL may vary among the eight different forms, which are distinguished typically on the basis of age at onset, clinical manifestations, and ultrastructural findings (Wisniewski et al., 1998, 1999, 2000). The typical ultrastructural findings of lysosomal storage of lipofuscin inclusions in NCL are composed of granular osmiophilic deposits (GRODs) in INCL, mainly curvilinear (CV) profiles in LINCL, and fingerprint (FP) profiles in JNCL (Wisniewski et al., 1998).

Eight genes, designated CLN_{1-8}, have been recognized to encode the different NCL forms. Five of them, CLN_{1-3}, CLN_5, and CLN_8, encoding INCL, LINCL, JNCL, FNCL (Finnish-variant LINCL), and progressive epilepsy with mental retardation (EPMR) respectively, have been cloned (International Batten Disease Consortium, 1995; Vesa et al., 1995; Sleat et al., 1997; Savukoski et al., 1998; Ranta et al., 1999). At least 77 mutations, including 26 in CLN_1, 26 in CLN_2, and 25 in CLN_3 (Mole et al., 1999), and four novel mutations, including three point mutations and one small deletion in CLN_2 (Zhong et al., unpublished data), have been identified in classical NCL-affected patients. Five mutations, the 223A→G (Y109D) and 451 C→T (R151X) in CLN_1, 636 C→T (R280X) and T523-1 G →C (IVS 5-1) in CLN_2, and a deletion of 1.02-kb genomic fragment involving exon 7-8 of CLN_3, have been characterized as common in INCL, LINCL, and JNCL (International Batten Disease Consortium, 1995; Munroe et al., 1997; Das et al., 1998; Zhong et al., 1998a, 1998b, 2000a). In addition, three mutations were observed in CLN_5 in which a 2-bp deletion in exon 4 accounts for almost 90% (17/19) of mutations detected (Savukoski et al., 1998). A missense mutation, 70C→G resulting in R24G, was found to be homozygous in EPMR patients and co-segregated with EPMR phenotypes (Ranta et al., 1999).

Molecular cloning, mutation identification in CLN genes, and characterization of common mutations in different forms of NCLs make it possible to offer molecular genetic testing for either confirming pathological findings and clinical diagnosis or identifying familial mutation(s) that would be beneficial for presymptom diagnosis, carrier screening, and prediction prenatally of outcome of pregnancy in NCL families. In this chapter, an overview of molecular genetic testing of NCL, along with diagnostic strategies for requesting genetic testing, is discussed.

II. SPECIMENS REQUIRED FOR GENETIC TESTING

A. Peripheral blood

Genomic DNA is the most common material used for genetic testing. The DNA can be extracted from any human tissues, including peripheral white blood cells. Usually, 5–10 ml of whole blood from an adult or 3–5 ml from a child are required for DNA extraction. The peripheral blood should be drawn into *purple-topped* vaccutainer tubes containing EDTA-K$_3$. Classically, genomic DNA is extracted from white blood cells with proteinase K and phenol/chloroform. This procedure provides good-quality DNA with little protein contamination, which can be used for both Southern hybridization and polymerase chain reaction (PCR). Recently, variant commercial kits have become available, most of which replace phenol/chloroform with high-salt solution to precipitate proteins (Buffone, 1985). In our laboratory, we use the Puregene kit (Gentra, Minneapolis, MN) for routine DNA extraction for NCL specimens. The quality and quantity of the isolated genomic DNA are good for both Southern hybridization and PCR and can be stored at 4°C for years without degradation.

RNA is used for research-based genetic testing. The RNA can be extracted directly from peripheral blood or more commonly from banked cell cultures of fibroblasts or lymphoblasts. For the latter 5–10 ml of blood are drawn in *green-topped* tubes, lymphocytes are isolated, and a long-term lymphoblast cell line is established by transfection with Epstein-Barr (EB) virus. The extracted RNA is used for Northern hybridization to detect gene deletion and mRNA splicing(s), and for reverse transcription (RT) to perform RT-PCR to study gene mutation(s) at the cDNA level.

B. Tissues

Genomic DNA can be extracted from any human tissues. Tissues may be obtained from biopsy or autopsy specimens, such as human brain, skin, or muscle. A commercial kit for isolating genomic DNA can be used. However, for extracting genomic DNA from brain tissues, we recommend to use the classical proteinase/phenol protocol.

III. MOLECULAR GENETIC TESTING FOR JNCL

A. Molecular testing for the 1.02-kb common deletion in the *CLN$_3$* gene

1. PCR amplification of the 1.02-kb common deletion

A 1.02-kb deletion of the genomic fragment involving exons 7 and 8 in the CLN$_3$ gene underlying JNCL was found to be the common mutation in JNCL alleles (International Batten Disease Consortium, 1995). DNA testing of this common deletion was described either with a protocol of long-range PCR (Zhong et al., 1998a) or a protocol of allele-specific amplification (Taschner et al., 1997). Both protocols can be scaled down to 10-µl reaction and detected by micro-agarose gel (Zhong et al., 2000).

Long-range PCR amplifies a single 3.3-kb genomic fragment from normal individuals, double bands of 2.3 kb and 3.3 kb from heterozygous carriers, and a single band of 2.3 kb from JNCL-affected patients (Zhong et al., 1998a). Allele-specific PCR detects a 729-bp single band from normal controls, 729-bp/426-bp bands from heterozygotes, and a 426-bp single band from JNCL-affected patients (Taschner et al., 1997). Both procedures have been validated for clinical molecular diagnostic testing (Zhong et al., 2000a).

2. Protocols of DNA testing for the 1.02-kb deletion

Primers used for detecting the common 1.02-kb deletion in the CLN_3 gene are provided in Table 7.1. PCR reaction is conducted in a final volume of 10 μl in 0.2-ml thin-wall PCR tube(s). Long-range PCR employs Taq Extender for amplifying long (>2-kb) fragments from genomic DNA (Table 7.2). Only the two-segment thermocycling condition (30 s at 94°C, and 5 min at 72°C) is performed. A regular PCR reaction containing three primers and routine condition with three-segment thermocycling is applied in allele-specific PCR (Table 7.3). PCR product is detected with 1% agarose gel (Zhong et al., 2000a).

Table 7.1. Primers Used for Testing Common Mutations in CLN_{1-3}

NCL	Gene	Mutation	Primer	Sequence	Ta
INCL[a]	CLN_1	223A→G	223F	ggagttgtaagtgagttatag	57°C
			223A	ccttacctccatcagggt	
			223G	ccttacctccatcagggg	
		451C→T	451F	ttctcacagtgccttgtgc	57°C
			451C	gctctctcctgggcatcg	
			451T	gctctctcctgggcatca	
LINCL[a]	CLN_2	IVS5-1	IntR1	ggtggtaaggaattgaggac	64°C
			T523-1G	agcctgacttctccctacag	
			T523-1C	agcctgacttctccctacac	
		636C→T	636C	ccctctgtgatccgtaagc	
			636T	ccctctgtgatccgtaagt	
JNCL	CLN_3	1.02-kb deletion	BDNZ1F[b]	atttctgtgggaccagcctgtgtg	72°C
			BDNZ2R	gcctcaggagatgtgagcaacaag	
			INTF7[c]	cattctgtcacccttagaagcc	56°C
			3.3R	ggggaggacaagcactg	
			2.3R	ggacttgaaggacggagtct	

Ta: annealing temperature for PCR-thermocycling.
[a]Zhong et al. (2000).
[b]Zhong et al. (1998a).
[c]Taschner et al. (1997).

Table 7.2. PCR Amplification for 1.02-kb Deletion in CLN_3

Reagents	Final concentration	1 × (μl)
Long-range PCR		
dH$_2$O	—	6.125
10 × Taq Extender buffer (Stratagene)	1×	1.0
dNTPs [10 mM] (Perkin-Elmer)	1 mM	1.0
Mixed primers (50 pmol/each)	0.3125 pmol	0.125
Taq Extender (Stratagene)	0.125 U/μl	0.5
Taq Polymerase (Perkin-Elmer)	0.125 U/μl	0.25
DNA	~1 ng/μl	1.0
Allele-specific PCR		
dH$_2$O	—	7.05
10 × Taq Buffer II (Perkin-Elmer)	1×	1.0
dNTPs [2.5 mM] (Perkin-Elmer)	0.2 mM	0.8
MgCl$_2$ [25 mM] (Perkin-Elmer)	1.5 mM	0.6
Mixed primers (50 pmol/each)	0.5 pmol	0.1
Taq Polymerase (Perkin-Elmer)	0.025 U/μl	0.05
DNA	~1 ng/μl	1.0

3. Interpretation of testing results

It has been reported that up to 96% of JNCL alleles carry the 1.02-kb deletion, which is most likely in a homogenous population, such as the Finnish (Järvelä et al., 1996), or in clinically typical and well-characterized JNCL patients (Munroe et al., 1997). However, testing for the 1.02-kb deletion in the CLN_3 gene may diagnose ~70% of clinically referred JNCL cases in North America (Zhong et al., 1998a, 2000). The remaining 30% would be uncommon mutations, including point mutations, insertions, and non-1.02-kb deletions (International Batten

Table 7.3. ASPE Reaction (μl) for Common Mutations in CLN_1 and CLN_2

	CLN_1	CLN_2
dH$_2$O	12.7	16.375
10 × Taq Buffer II (Perkin-Elmer)	2.0	2.5
dNTPs [2.5 mM] (Perkin-Elmer)	1.6	2.0
MgCl$_2$ [25 mM] (Perkin-Elmer)	1.2	1.25
Common primer [50 pmol/μl]	0.2	0.125
Allele-specifc primer [50 pmol/μl]	0.2	0.125
Taq Polymerase [5u/μl] (Perkin-Elmer)	0.1	0.125
DNA [10 ng/μl]	2.0	2.5
Total	20	25

Disease Consortium, 1995; Munroe et al., 1997; Zhong et al., 1998a). Three possible DNA testing results: a single normal band "N," a single mutant band "M," or double-bands of both normal and mutant "N/M," are expected.

1. Interpretation of finding of a single PCR band "M" is that the subject is affected with JNCL.
2. Detection of both "N" and "M" bands indicates that the subject is heterozygous for the 1.02-kb deletion. The subject could be either an affected JNCL patient who carries another uncommon mutation on the second allele or a carrier who does not have a mutation on the second chromosome. Distinguishing JNCL-affected patients from carriers requires integration of clinical information and pathological findings. Usually, this is not difficult to obtain for elderly individuals once clinical neurodegenerative symptoms and electron microscopic (EM) fingerprint (FP) profiles are documented. However, it is not easy for younger siblings who have not yet reached the age of clinic onset.
3. Interpretation of the finding of the homozygote of "N" alleles is complicated. A single "N" band can be seen in normal individuals from the general population as well as in JNCL-affected individuals who have uncommon mutations on both chromosomes. If a subject has typical JNCL clinical and pathological phenotypes but no 1.02-kb deletion has been detected on both chromosomes, further study with gene scan on the CLN_3 locus to investigate mutation(s) other than the 1.02-kb deletion is highly recommended.

B. Molecular detection of uncommon mutations

1. Single-strand conformation polymorphism (SSCP) analyses

To date, at least 24 mutations have been reported as familial-specific or "private" mutations in the CLN_3 gene (Mole et al., 1999). They spread along the CLN_3 locus (Munroe et al., 1997) that encodes a 1.7-kb transcript (International Batten Disease Consortium, 1995) and spans 16 kb in the genomic structure (Mitchison et al., 1998a). These mutations include point mutations and large and small deletions. Some of them involve intron–exon junction sites that lead to splicing errors. Most mutations that account for 63% (15/24) cluster at exons 6, 7, 8, and 13. So far, no mutation has been reported at exons 1 to 4 or at the promoter region. Application of SSCP technology combined with direct DNA sequencing can specifically analyze the clustered regions and possibly detect the clustered mutations (Munroe et al., 1997; Zhong et al., 1998b); this methodology is sensitive enough, combined with the sensitivity of detecting the 1.02-kb deletion, of detecting ~90% of JNCL mutations reported.

2. RT-PCR to analyze mutation in the CLN_3 coding region

One approach for detecting any mutation(s) in the cDNA region, including aberrant splicing, is reverse-transcribing mRNA into cDNA, followed by PCR to amplify the CLN_3 coding region. Total RNA extracted, as described above, from cultured lymphcytes or fibroblasts, or directly from whole blood, is used as template for reverse transcription (RT). The synthesized cDNA is amplified by primers (forward: 5' gacctgaacttgatg, reverse: 5' ctgcgtcctgaggatcc) to obtain a 1.3-kb fragment of the CLN_3 coding sequence. This approach is good for detecting deletion/insertion or aberrant splicing when the amplified PCR products are mobilized in high-resolution gel. However, we have found that RT-PCR amplified mRNA products were heterogenous species, which included a lot of *in vivo* alternatively spliced mRNA molecules with unknown reason, especially when long-term cultures of EB-tranfected lymphoblasts were used as the source of extracting RNA (Zhong et al., unpublished data). Employing cultured fibroblasts or RNA extract directly from fresh blood may improve the artificial splicing forms of CLN_3 mRNA, but once alternatively spliced mRNA is detected, it is difficult to interpret the result unless mutation(s) can be confirmed from intron–exon junction(s) in genomic DNA.

3. Gene scan for CLN_3 locus

Gene scan refers to molecular studies of the whole CLN_3 gene locus. Technically, it refers to PCR amplification and sequencing analyses of the whole genomic sequence of the CLN_3 genomic structure, which would include the whole exon sequences, intron–exon junctions, and CLN_3 promoters as well as some intron sequences. We have developed PCR/sequencing analyses to conduct gene scans that analyze all exons, intron–exon junctions, and ~1-kb of promoter region sequences (Zhong et al., unpublished data). The amplified genomic DNA fragments ranges from 0.5 to 1.2 kb. Direct sequencing of the PCR-amplified DNA fragments would allow detecting any mutation(s) embedded.

Gene scan is labor-intensive, costly, and time-consuming. Usually, it is done on a research-based analysis. Applying robotic automation may stimulate this study. However, gene scan may be the best approach for searching for mutations in JNCL-affected patients who have been characterized to be heterozygous for the 1.02-kb deletion but for whom a mutation on the second allele is unknown. In addition, gene scan may be applied to study clinically typical JNCL cases with typical fingerprint (FP) profile detected pathologically but for which no mutations could be detected by either SSCP or RT-PCR. Generally speaking, if no mutation can be found by gene scan, a variant form of NCL should be considered and other loci such as CLN_1 should be studied (see discussion that follows).

4. Southern hybridization for detecting large deletion

Large deletions involving the CLN_3 locus may not be detected by the molecular technology described above, especially if the deletion involves only one chromosome. Four deletions, including the 1.02-kb common deletion, have been reported, among which the largest deletion is 6 kb, spanning intron 5 to intron 15. The three uncommon deletions account for 12.5% (3/24) of CLN_3 mutations. Use of a probe of cDNA2-3 to hybridize genomic DNA on Southern blot that is digested with restriction enzyme *Pst I* may detect these deletions (International Batten Disease Consortium, 1995).

IV. MOLECULAR GENETIC TESTING FOR LINCL AND INCL

A. Molecular testing for common mutations in INCL and LINCL

1. Amplification of common mutations in CLN_1 and CLN_2

Four point mutations, 223A→G (Y109D) and 451 C→T (R151X) in CLN_1, and 636 C→T (R280X) and T523-1 G→C (IVS 5-1) in CLN_2, have been characterized as common (Das *et al.*, 1998; Zhong *et al.*, 1998b) and account for two-thirds of clinically referred INCL and LINCL cases (Zhong *et al.*, 2000a). PCR protocols with allele-specific primer extension (ASPE) have been applied successfully in amplifying specifically for these four common mutations (Zhong *et al.*, 2000a).

2. ASPE-PCR

ASPE employs two sets of PCR reactions, one for detecting the normal (N) allele and the other for the mutant (M) allele, to test a specific point mutation. Each reaction contains a common primer and an allele-specific primer (Table 7.1). PCR is conducted in 20- to 25-µl reactions (Table 7.3), and four controls, including one normal, one heterozygote, and one homozygote of the specific mutation and one blank with dH_2O, are always included in each batch of ASPE amplification (Fig. 7.1).

3. Interpretation of test results

The three outcomes obtained from ASPE-PCR consist of N/N, N/M, and M/M. The N presents a normal allele, and M presents a mutant allele (Figure 7.1).
1. A DNA testing result of N/N indicates that only one normal allele could be detected from both chromosomes. This is always seen in normal controls and

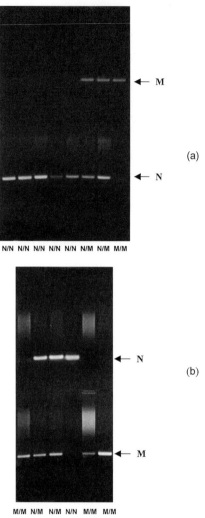

Figure 7.1. ASPE-PCR of common mutations in CLN_1 and CLN_2. Arrows indicate normal (N) and mutant (M) alleles. Testing for mutation 451T→C (R151X) in CLN_1 (a), and 636 C→T (R280X) in CLN_2 (b) is shown. M/M is homozygous and N/M is heterozygous for the mutation. N/N is homozygous with no mutation detected. Lane "B" is a blank control.

in NCL-affected individuals who do not carry the specific mutation being tested in both chromosomes.

2. A result of N/M indicates that the subject being tested carries the specific mutation in one chromosome, but not in the second chromosome. This is seen either in carriers such as parents and normal siblings of an affected child in an NCL family. Affected individuals who have a common mutation that is

being tested in one allele but another uncommon mutation in the second allele may also show the N/M pattern.
3. M/M is interpreted as NCL-affected because mutations are found on both chromosomes. The individual is homozygous for the mutation.

4. Alternative analyses of common mutations in CLN_2

PCR amplification of a fragment that contains both mutations 636 C→T (R280X) and T523-1 G→C (IVS 5-1) in CLN_2 followed by direct DNA sequencing analyses has been described previously (Zhong et al., 1998b). Using this approach, we identified a novel mutation at IVS5-1, T523-1 G→A, causing aberrant RNA splicing (Hartikainen et al., 1999).

B. Gene scan of uncommon mutations in LINCL

1. Mutations identified from CLN_2 locus

The genomic structure of CLN_2 locus, which is localized at 11p15, consists of 13 exons and 12 introns and spans a 6.7-kb genomic sequence (Liu et al., 1998). A total of 26 mutations have been identified in the CLN_2 gene (Mole et al., 1999). Using 13 pairs of primers, Sleat et al. (1999) identified 24 mutations, including the two common mutations as well as T523-1 G→A described above from 74 classical LINCL families. Unlike mutations for JNCL that cluster at certain exons in CLN_3, mutations in CLN_2 spread along the whole coding sequence (Sleat et al., 1999).

2. Gene scan of CLN_2

We have developed a gene scan approach to study the whole CLN_2 locus (GenBank Accession # AF039704). Four genomic fragments ranging from 0.9 to 1.3 kb that include the entire CLN_2 coding sequence, all exon–intron junctions, some intronic sequences, and a ~500-bp promoter region (Liu et al., 1998) can be PCR-amplified and sequenced. Recently, four novel mutations, two missense mutations (G284V, R127Q), one nonsense mutation (W460X), and one small deletion, have been identified (Zhong et al., 2000b) in LINCL-affected individuals who were heterozygous for one of the two common mutations.

V. MOLECULAR SCREENING OF CARRIER STATUS IN NCL FAMILIES

A. Family members of individuals with NCL are encouraged to be screened for carrier status

NCLs encoded by CLN_{1-8} genes are inherited in an autosomal recessive manner. Parents of NCL-affected individuals are obligate carriers. When mutations identified from the proband who is affected with NCL are not identical on the two chromosomes, characterization of the genetic inheritance of mutations should be the primary genetic testing performed for the family in order to obtain information about which mutation derives from which chromosome. The possibility of being a carrier for nonsymptomatic sibling(s) of the proband is 2/3 and of being NCL-affected is 1/3. Genetic testing for these sibling(s) may help identify carrier status and make presymptomatic diagnosis. Grandparents from both maternal and paternal sides should be tested to trace the mutation. Secondary-degree relatives of the proband, such as uncles and aunts, are encouraged to have genetic testing because they all have a 50% possibility of being carriers. If carrier status is confirmed, their offspring can be offered the test.

B. Characterize familial mutations before screening for carrier status is requested

Information on familial mutation(s) should be known in advance of requesting molecular testing for screening carrier status. This information may help the DNA diagnostic laboratory study the specific locus. If there is no familial information regarding the mutation(s), a nonspecific "shotgun" approach to exclude common mutations may help identify genetic risk of being a carrier (Zhong et al., 2000a). Usually, this nonspecific testing is used only for the normal mate of an NCL carrier to predict genetic risk for their offspring(s).

We have performed carrier screening in 165 individuals from genetically characterized NCL families, among which we identified 70 (42%) carriers (Zhong et al., 2000a).

VI. PRENATAL DIAGNOSTIC TESTING FOR NCL

A. Specimens used for prenatal molecular testing

Chorionic villus samples (CVS) and aminiotic fluid (AF) obtained during the first trimester are the most common specimens used for routine prenatal testing, although fetal blood cells can be isolated from maternal peripheral blood at an earlier stage. Ten to twenty milligrams (mg) of CVS is usually obtained by 10–12 weeks' gestation, but could be obtained as early as the ninth gestational week. CVS specimens must be dissected to clean up maternal materials contaminated before sending to the DNA diagnostic laboratory. AF is usually obtained by 14–16 weeks' gestation.

DNA can be isolated directly from CVS and AF cells, although the yield of recovered DNA is low. Directly isolated DNA is usually enough for most PCR-based molecular testing, which can shorten the period of obtaining testing results.

Cultured CVS or AF is preferred to confirm genetic testing results obtained from directly extracted DNA and kept as backup specimens. This is very important if the pregnant mother is a heterozygote and testing result showed that the fetus is also heterozygous. Cultured CVS or AF must be tested to confirm that the carrier status of the fetus is not an artificial result from the contamination of maternal materials.

Other fetal tissues including fetal blood, skin, liver, umbilical vessels, and brain can be used for prenatal diagnosis, too (Chow et al., 1993). Prenatal diagnosis may also be performed at an earlier stage in single blastomeres from in vitro-fertilized embryos (Syvanen et al., 1997).

B. Information about familial mutation(s) is a basic requirement for prenatal molecular genetic testing

To offer prenatal diagnostic testing, basic information about familial mutation(s) carrier status should be obtained in advance. Once the familial mutations have been characterized, molecular genetic testing will be able to detect the specific mutations for fetus.

Information about prenatal testing may provide accurate genetic counseling. If no information regarding the familial mutation(s) is available when prenatal testing is requested, the result of prenatal testing for common mutation(s) in pregnant women should be integrated with Bayesian calculation when genetic counseling is provided to evaluate genetic risk (Zhong et al., 2000).

C. Molecular approaches for prenatal diagnosis of NCL

Before the era of molecular genetics, prenatal diagnosis of NCL depended on ultrastructural studies with electron microscopy (EM). MacLeod et al. (1984, 1985, 1988) successfully identified curvilinear (CV) inclusions from uncultured AF cells at 16 weeks' gestation, by which LINCL was diagnosed. Examination of uncultured CVS tissues, with either membrane-bound inclusions in typical INCL (Rapola et al., 1988, 1990; Goebel et al., 1995) or fingerprint (FP) inclusions in typical JNCL (Conradi et al., 1989) could be observed in the first trimester of pregnancies. Negative evidence of FP in uncultured AF cells, lymphocytes isolated from fetal blood, or fetal skin biopsy specimens excludes the presence of JNCL (Kohlschutter et al., 1989).

If the gene underlying a particular NCL has not yet been cloned, *indirect testing* with linkage studies would play an important role. Molecular-based prenatal diagnosis was first applied for INCL with restriction fragment length polymorphism (RFLP) (Järvelä et al., 1991). PCR-based polymorphic marker may also be employed for *indirect* prenatal diagnosis, once the underlying gene is assigned to a particular locus (Vesa et al., 1993; Taschner et al., 1995).

Presently, most molecular prenatal diagnoses are conducted with *direct* molecular testing, for which information about familial mutation(s) is valuable to focus the prenatal testing with the molecular technology described above. The first DNA-based *direct* prenatal diagnosis was reported by Munroe et al. (1996) to test a 1.02-kb deletion in a JNCL-affected family. Currently, molecular testing of specific mutation(s) is available for INCL (de Vries et al., 1999), LINCL (Berry-Kravis et al., 2000), JNCL (Munroe et al., 1996), and FNCL (Syvanen et al., 1997; Rapola et al., 1999). Theoretically, it should be available for EPMR, too, because most EPMR contains a common missense mutation of 70C→G (Ranta et al., 1999).

Recently, our laboratory has conducted molecular prenatal tests for five pregnancies with uncultured AF or CVS, one for INCL, three for LINCL, and one for JNCL (Zhong et al., unpublished data).

VII. IMPORTANT ISSUES IN THE MOLECULAR GENETIC TESTING

A. Testing five common mutations in CLN_1, CLN_2, and CLN_3 may detect 70% of classical INCL, LINCL, and JNCL cases

Testing for the five common mutations, the 223A→G and 451 C→T in CLN_1, 636 C→T and T523-1 G→C in CLN_2, and a 1.02-kb deletion in CLN_3, allows identification of at least 70% of classical pediatric NCL cases (Zhong et al., 2000). This figure, obtained from testing 180 NCL families (Table 7.4), confirmed that these five mutations account for the majority of NCL mutations. However, negative results from testing the common mutations should be carefully interpreted and detailed explanation should be provided for patients and families who request genetic testing, because the subject who has been tested may still carry uncommon mutation(s) that account for the remaining 30% of clinically referred NCL cases.

Table 7.4. Frequency of NCL Homozygotes and Heterozygotes

	Frequency (%)		
Mutations	INCL (n = 27)	LINCL (n = 76)	JNCL (n = 77)
Homozygotes	6 (22.2)	14 (18.4)	42 (54.5)
Double hetrozygotes	4 (14.8)	7 (9.2)	
Single hetrozygotes	11 (40.7)	33 (43.4)	16 (24.7)
Total probands detected	21 (77.8)	54 (65.8)	58 (75.3)

B. Clinical and pathological data should be collected before genetic testing is requested

Clinical symptoms, such as the age at onset, the first symptom of seizure or vision loss, and pathological findings, are important to evaluate different forms of NCL (Wisniewiski et al., 1992, 1998, 1999). Ultrastructural study should be conducted before the case is referred to DNA testing. Classical NCLs usually show typical EM profile(s), which may help in determining which CLN gene should be the focus of initial DNA testing. The importance of pathological findings has been documented in atypical JNCL and LINCL patients who carry GROD profiles (Mitchison et al., 1998b); in these patients, mutations on the CLN_1 gene, rather than CLN_2 or CLN_3, should be tested initially. If EM cannot detect lysosomal inclusions of lipofuscin storage from buffy coats or skin punch biopsy, the referred individual is usually not likely to have NCL. If INCL or LINCL can be determined by testing PPT1 or TPP1 (Vesa et al., 1995; Sleat et al., 1997) to show lysosomal enzyme deficiency, gene scan should be offered definitely to characterize familial mutation(s).

Occasionally, in clinically and pathologically diagnosed NCL patient(s), deficiency cannot be documented in the CLN_{1-3} genes that underlie classical NCLs; in these cases, the gene or locus for a variant form of NCL should be considered. Ethnic background is as an important issue to direct DNA testing for CLN_{5-8}, although CLN_{6-7} are still being identified and cloned.

C. Molecular screening of carrier status in NCL-affected families is encouraged

As a result of genetic testing, the incidence of NCL is now higher than it used to be. The estimated NCL frequency has increased from 1/100,000 to 1/25,000–1/12,500 (Rider and Rider, 1997; Mole, 1998), and the carrier frequency has increased from 1/150 to 1/55–1/80 in the general population. If a carrier mates randomly with an individual from the general population, the possibility of having an affected child in the family would be 1/220 to 1/320. DNA testing will help predict more accurately the genetic risks for the pregnancy. If negative results were obtained by testing for the five common mutations, which detects ~70% of NCL mutations, the risk of the couple having an affected infant would be reduced to ~1/400 to 1/740 (Zhong et al., 2000a).

Currently, no treatment is available for the NCLs. Knowing carrier status and having prenatal testing is the only way to prevent NCL-affected pregnancies. Because of the increased frequency of both affected individuals and carriers, molecular screening in children who have progressive seizures or impairment of vision with no other abnormalities observed has been proposed (Zhong et al., 2000a), which may recognize and identify NCL patients and families at an earlier stage.

Acknowledgments

This study was supported in part by grants from the Batten Disease Support and Research Foundation (BDSRA), the New York State Office of Mental Retardation and Developmental Disabilities (OMRDD), and the Children's Brain Research Foundation. We would like to thank all the families and physicians who have participated in our research studies.

References

Bennett, M. J., and Hofmann, S. L. (1999). The neuronal ceroid lipofuscinoses (Batten disease): A new class of lysosomal storage diseases. *J. Inherit. Metab. Dis.* **22**, 535–544.
Berry-Kravis, E., Sleat, D. E., Sohar, I., Meyer, P., Donnelly, R., and Lobel, P. (2000). Prenatal testing for late infantile neuronal ceroid lipofuscinosis. *Ann. Neurol.* **47**, 254–257.
Buffone, G. J. (1985). Isolation of DNA from biological specimens without extraction with phenol. *Clin. Chem.* **31**, 164–165.
Chow, C. W., Borg, J., Billson, V. R., and Lake, B. D. (1993). Fetal tissue involvement in the late infantile type of neuronal ceroid lipofuscinosis. *Prenat. Diagn.* **13**, 833–841.
Conradi, N. G., Uvebrant, P., Hokegard, K. H., Wahlstrom, J., and Mellqvist, L. (1989). First-trimester diagnosis of juvenile neuronal ceroid lipofuscinosis by demonstration of fingerprint inclusions in chorionic villi. *Prenat. Diagn.* **9**, 283–287.
Das, A. K., Becerra, C. H. R., Yi, W., Lu, J-Y., Siakotos, A. N., Wisniewski, K. E., and Hofmann, S. L. (1998). Molecular genetics of palmitoyl-protein thioesterase deficiency in the U.S. *J. Clin. Invest.* **102**, 361–370.
de Vries, B. B., Kleijer, W. J., Keulemans, J. L., Voznyi, Y. V., Franken, P. F., Eurlings, M. C., Galjaard, R. J., Losekoot, M., Catsman-Berrevoets, C. E., Breuning, M. H., Taschner, P. E., and van Diggelen, O. P. (1999). First-trimester diagnosis of infantile neuronal ceroid lipofuscinosis (INCL) using PPT enzyme assay and CLN1 mutation analysis. *Prenat. Diagn.* **19**, 559–562.
Goebel, H. H., Vesa, J., Reitter, B., Goecke, T. O., Schneider-Ratzke, B., and Merz, E. (1995). Prenatal diagnosis of infantile neuronal ceroid-lipofuscinosis: A combined electron microscopic and molecular genetic approach. *Brain Dev.* **17**, 83–88.
Hartikainen, J. M., Ju, W., Wisniewski, K. E., Moroziewicz, D. N., McLendon, L., Zhong, D., Brown, W. T., and Zhong, N. (1999). Late infantile neuronal ceroid lipofuscinosis is due to splicing mutations in the CLN2 gene. *Mol. Genet. Metab.* **67**, 162–168.
International Batten Disease Consortium (1995). Isolation of a novel gene underlying Batten disease, CLN3. *Cell* **82**, 949–957.
Järvelä, I., Rapola, J., Peltonen, L., Puhakka, L., Vesa, J., and Ammala, P., *et al.* (1991). DNA-based prenatal diagnosis of the infantile form of neuronal ceroid lipofuscinosis (INCL, CLN1). *Prenat. Diagn.* **11**, 323–328.
Järvelä, I., Mitchison, H. M., Munroe, P. B., O'Rawe, A. M., Mole, S. E., and Syvänen, A-C. (1996). Rapid diagnostic test for the major mutation underlying Batten disease. *J. Med. Genet.* **33**, 1041–1042.
Kohlschutter, A., Rauskolb, R., Goebel, H. H., Anton-Lamprecht, I., Albrecht, R., and Klein, H. (1989). Probable exclusion of juvenile neuronal ceroid lipofuscinosis in a fetus at risk: An interim report. *Prenat. Diagn.* **9**, 289–292.
Liu, C-G., Sleat, D. E., Donnelly, R. J., and Lobel, P. (1998). Structural organization and sequence of CLN2, the defective gene in classical late infantile neuronal ceroid lipofuscinosis. *Genomics* **50**, 206–212.
MacLeod, P. M., Dolman, C. L., Nickel, R. E., Chang, E., Zonana, J., and Silvey, K. (1984). Prenatal diagnosis of neuronal ceroid lipofuscinosis. *N. Engl. J. Med.* **310**, 595.

MacLeod, P. M., Dolman, C. L., Nickel, R. E., Chang, E., Nag, S., Zonana, J., and Silvey, K. (1985). Prenatal diagnosis of neuronal ceroid-lipofuscinoses. *Am. J. Med. Genet.* **22,** 781–789.

MacLeod, P. M., Nag, S., and Berry, C. (1988). Ultrastructural studies as a method of prenatal diagnosis of neuronal ceroid-lipofuscinosis. *Am. J. Med. Genet.* **5,** 93–97.

Mitchison, H. M., Munroe, P. B., O'Rawe, A. M., Taschner, P. E. M., de Vos, N., and Kremmidiotis, G., *et al.* (1998a). Genomic structure and complete nucleotide sequence of the Batten disease gene, CLN3. *Genomics* **40,** 346–350.

Mitchison, H. M., Hofmann, S. L., Becerra, C. H. R., Munroe, P. B., Lake, B. D., and Crow, Y. J., *et al.* (1998b). Mutations in the palmitoyl-protein thioesterase gene (PPT; CLN1) causing juvenile neuronal ceroid lipofuscinosis with granular osmiophilic deposits. *Hum. Mol. Genet.* **7,** 291–297.

Mole, S. E. (1998). Batten disease: Four genes and still counting. *Neurobiol. Dis.* **5,** 287–303.

Mole, S. E., Mitchison, H. M., and Munroe, P. B. (1999). Molecular basis of the neuronal ceroid lipofuscinoses: Mutations in CLN1, CLN2, CLN3, and CLN5. *Hum. Mutat.* **14,** 199–215.

Munroe, P. B., Rapola, J., Mitchison, H. M., Mustonen, A., Mole, S. E., Gardiner, R. M., and Jarvela, I. (1996). Prenatal diagnosis of Batten's disease. *Lancet* **347,** 1014–1015.

Munroe, P. B., Mitchison, H. M., O'Rawe, A. M., Anderson, J. W., Boustany, R.-M., and Lerner, T. J., *et al.* (1997). Spectrum of mutations in the Batten disease gene, CLN3. *Am. J. Hum. Genet.* **61,** 310–316.

Ranta, S., Zhang, Y., Ross, B., Lonka, L., Takkunen, E., and Messer, A., *et al.* (1999). The neuronal ceroid lipofuscinoses in human EPMR and mnd mutant mice are associated with mutations in CLN8. *Nat. Genet.* **23,** 233–236.

Rapola, J., Santavuori, P., and Heiskala, H. (1988). Placental pathology and prenatal diagnosis of infantile type of neuronal ceroid-lipofuscinosis. *Am. J. Med. Genet.* **5,** 99–103.

Rapola, J., Salonen, R., Ammala, P., and Santavuori, P. (1990). Prenatal diagnosis of the infantile type of neuronal ceroid lipofuscinosis by electron microscopic investigation of human chorionic villi. *Prenat. Diagn.* **10,** 553–559.

Rapola, J., Lahdetie, J., Isosomppi, J., Helminen, P., Penttinen, M., and Jarvela, I. Prenatal diagnosis of variant late infantile neuronal ceroid lipofuscinosis (vLINCL[Finnish]; CLN5). (1999) *Prenat. Diagn.* **19,** 685–688.

Rider, J. A., and Rider, D. L. (1997). Batten disease, a twenty-eight-year struggle: Past, present and future. *Neuropediatrics* **28,** 4–5.

Savukoski, M., Klockers, T., Holmberg, V., Santavuori, P., Lander, E. S., and Peltonen, L. (1998). CLN5, a novel gene encoding a putative transmembrane protein mutated in Finnish variant late infantile neuronal ceroid lipofuscinosis. *Nat. Genet.* **19,** 286–288.

Sleat, D. E., Donnelly, R. J., Lackland, H., Liu, C-G., Sohar, I., Pullarkat, R. K., and Lobel, P. (1997). Association of mutations in a lysosomal protein with classical late-infantile neuronal ceroid lipofuscinosis. *Science* **277,** 1802–1805.

Sleat, D. E., Gin, R. M., Sohar, I., Wisniewski, K., Sklower-Brooks, S., and Pullarkat, R. K., *et al.* (1999). Mutational analysis of the defective protease in classic late-infantile neuronal ceroid lipofuscinosis, neurodegenerative lysosomal storage disorder. *Am. J. Hum. Genet.* **64,** 1511–1523.

Syvanen, A. C., Järvelä, I., and Vesa, J. (1997). DNA diagnosis and identification of carriers of infantile and juvenile neuronal ceroid lipofuscinoses. *Neuropediatrics* **28,** 63–66.

Taschner, P. E. M., de Vos, N., Post, J. G., Meijers-Heijboer, E. J., Hofmann, I., and Loonen, M. C., *et al.* (1995). Carrier detection of Batten disease (juvenile neuronal ceroid-lipofuscinosis). *Am. J. Med. Genet.* **57,** 333–337.

Taschner, P. E. M., de Vos, N., and Breuning, M. H. (1997). Rapid detection of the major deletion in the Batten disease gene CLN3 by allele specific PCR. *J. Med. Genet.* **34,** 955–956.

Vesa, J., Hellsten, E., Makela, T. P., Jarvela, I., Airaksinen, T., Santavuori, P., and Peltonen, L. (1993). A single PCR marker in strong allelic association with the infantile form of neuronal ceroid lipofuscinosis facilitates reliable prenatal diagnostics and disease carrier identification. *Eur. J. Hum. Genet.* **1,** 125–132.

Vesa, J., Hellsten, E., Verkruyse, L. A., Camp, L. A., Rapola, J., Santavuori, P., Hofmann, S. L., and Peltonen, L. (1995). Mutations in the palmitoyl protein thioesterase gene causing infantile neuronal ceroid lipofuscinosis. *Nature* **376,** 584–587.

Wisniewski, K. E., Kida, E., Patxot, O. F., and Connell, F. (1992). Variability in the clinical and pathological finding in the neuronal ceroid lipofuscinoses: Review of data and observations. *Am. J. Med. Genet.* **42,** 525–532.

Wisniewski, K. E., Zhong, N., Kaczmarski, W., Lach, A., Sklower-Brooks, S., and Brown, W. T. (1998). Studies of atypical JNCL suggest overlapping with other NCL forms. *Pediatr. Neurol.* **18,** 36–40.

Wisniewski, K. E., Kaczmarski, A., Kida, E., Connell, F., Kaczmarski, W., Michalewski, M. P., Woroziewicz, D. N., and Zhong, N. (1999). Reevaluation of neuronal ceroid lipofuscinoses: Atypical juvenile onset may be the result of CLN2 mutations. *Mol. Genet. Metab.* **66,** 248–252.

Zhong, N. (2000). Neuronal ceroid lipofuscinoses and possible pathogenic mechanism. *Mol. Genet. Metab.* **71,** 195–206.

Zhong, N., Moroziewicz, D. N., Jurkiewicz, A., Ju, W., Johnston, L., Wisniewski, K. E., and Brown, W. T. (2000b). Heterogeneity of late-infantile neuronal ceroid lipofuscinosis (LINCL). *Genet. in Med.* **2(6),** in press.

Zhong, N., Wisniewski, K. E., Hartikainen, J., Ju, W., Moroziewicz, D. N., Mclendon, L., Sklower Brooks, S., and Brown, W. T. (1998a). Two common mutations in the CLN2 gene underlie late infantile neuronal ceroid lipofuscinosis. *Clin. Genet.* **54,** 234–238.

Zhong, N., Wisniewski, K. E., Kaczmarski, A. L., Ju, W., Xu, W. M., Xu, W. W., Mclendon, L., Liu, B., Kaczmarski, W., Sklower-Brooks, S., and Brown, W. T. (1998b). Molecular screening of Batten disease: Identification of a novel missense mutation (E295K). *Hum. Genet.* **102,** 57–62.

Zhong, N., Wisniewski, K. E., Ju, W., Moroziewicz, D. N., Jurkiewicz, A., Jenkins, E. C., and Brown, W. T. (2000a). Molecular diagnosis of, and carrier screening for, the neuronal ceroid lipofuscinoses (NCLs). *Genet. Testing* **4,** 243–248.

8

Genetic Counseling in the Neuronal Ceroid Lipofuscinoses

Susan Sklower Brooks[*]
Department of Human Genetics
New York State Institute for Basic Research in Developmental Disabilities
Staten Island, New York 10314

   I. Introduction
  II. Inheritance
 III. Genetics
  IV. Diagnostic Confirmation
   V. Genetic Counseling
  VI. Carrier Screening
 VII. Reproductive Options
VIII. Conclusion
       References

I. INTRODUCTION

The neuronal ceroid lipofuscinoses (NCLs), commonly known as Batten disease, are the most frequent neurodegenerative disorders of childhood. In the past their diagnosis was based entirely on clinical and pathological findings. Today, advances in our understanding of these disorders on a biochemical and molecular level has led to definitive testing suitable for diagnosis of affected individuals, presymptomatic diagnosis, and carrier identification. This chapter reviews genetic counseling for the NCLs.

[*]Address for correspondence: E-mail: Susan.Brooks@omr.state.ny

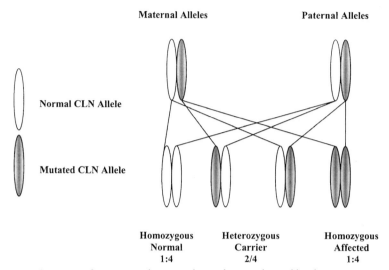

Figure 8.1. Segregation of recessive forms of neuronal ceroid lipofuscinosis.

II. INHERITANCE

All of the NCLs are inherited in an autosomal recessive manner with the exception of a very rare adult type (MIM163350, Parry disease) that is dominant. For the autosomal recessive disorders, each parent of an affected child is an obligate carrier and has a 50/50 chance of passing on the mutated gene. Thus, as shown in Figure 8.1, in any given pregnancy the risk that such an obligate heterozygous carrier couple will have affected child is 25% (1/4); 50% (1/2) for the child to be a carrier; and 25% (1/4) chance for the child to inherit two normal alleles. For each unaffected child the chance that he or she is a carrier is 66% (2/3).

III. GENETICS

While four discrete clinical syndromes were initially defined based on the age of onset and pathology, it has been clear for many years that there is genetic heterogeneity in the NCLs (see Chapter 1). This heterogeneity is due to both different genetic loci and multiple mutations within each specific locus.

Molecular dissection of the NCLs has confirmed that there are at least eight distinct genetic loci (Mole, 1999). They are designated CLN1 through CLN8. Six of these loci have been mapped and each is on a different chromosome (see Table 1.2; Järvelä et al., 1991a; Sharp et al., 1997; Gardiner et al., 1990;

Savukoski *et al.*, 1998; Tahvanainen *et al.*, 1994). Within each distinct locus, many disease causing mutations have been identified that explain the allelic heterogeneity (www.ucl.ac.uk/ncl).

IV. DIAGNOSTIC CONFIRMATION

Diagnostic confirmation is critical to genetic counseling. The NCLs must be distinguished from other neurodegenerative disorders of childhood. Before genetic counseling, all records must be reviewed with careful assessment of the clinical phenotype and morphology (see Chapters 1 and 2). The family history should be carefully detailed in a pedigree. The proband, if available, should be examined. Definitive laboratory testing should be performed to confirm the clinical diagnosis. If the proband is not available, testing could be carried out on the parents.

Recent developments in molecular and biochemical research has led to the development of several reliable diagnostic tests to confirm the diagnosis and classify the subtypes (see Chapters 4 and 7).

V. GENETIC COUNSELING

Once this information is obtained, the counselor can proceed with discussion of the issues and risks with the family. For many of the NCLs the onset is late and the diagnosis may take several years to be recognized. With the diagnosis, however, the devastating nature of these conditions becomes clear to the family. By the time the diagnosis of NCL is made in the proband, the family has often had several children. When the genetic counselor begins the discussion, it is not uncommon for the family's concern to shift from the proband to the younger children in the family, as the parents suddenly realize that their younger children are also at risk to be affected.

The genetic counselor needs to address many psychosocial issues, some of which are common to families of handicapped children in general and some that relate to the progressive nature of the diseases. Issues that may arise are the anxiety provoked by not knowing the diagnosis initially and then the trauma of learning the progressive nature and fatal prognosis. Parents may have feelings of frustration, because treatment is limited to symptomatic therapy. They grieve for the "normal" child they had hoped for and lost. Parents often experience a sense of isolation from their friends and family.

Several NCL support groups have formed that empower parents and reduce isolation. The Batten Disease Support and Research Association holds

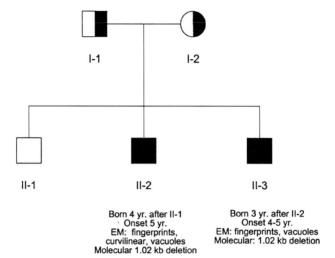

Figure 8.2. Pedigree of a family with juvenile neuronal ceroid lipofuscinosis. All three children were born prior to the onset of symptoms in the proband.

annual meetings for families and professionals, issues a newsletter, and provides telephone and Internet contacts. The Children's Brain Disease Research Foundation has been instrumental in lobbying Congress for funding for research in NCL and for supporting meetings where families and researchers gather to advance the understanding of the NCLs. The Institute for Basic Research Batten Disease Registry collects clinical, pathological, and genetic data from families to aid in research studies. In addition, the Registry provides expertise to assist in the diagnosis of atypical cases. The Internet has also provided an invaluable resource. Several Web pages relevant to NCL are listed in Table 8.1.

VI. CARRIER SCREENING

Families also have great concern that their unaffected children may be carriers. Although the gene frequency in most populations is relatively small, which minimizes the risk that unrelated spouses will be carriers, the parents realize that they, by chance, are both carriers. Thus, their concern for their unaffected children is high. There are several concerns regarding the timing for testing of unaffected children and adolescents (Wertz et al., 1994). The children may be overwhelmed with their sibling's illness and not ready to understand or cope with testing (Fanos, 1997). The testing may show that the individual is affected but presymptomatic.

Table 8.1. Useful NCL Addresses and Websites

Sponsor	Website	Information
Batten Disease Support and Research Association 2600 Parsons Avenue Columbus, Ohio 43207-2972 Tel. 800-448-4570	www.bdsra.org	Information about the diseases; links to other sites; bulletin board
Children's Brain Disease Foundation 350 Parnassas Avenue, Suite 900 San Francisco, CA 94117 Tel. 415-566-5402		Provides information to the public, raises public awareness, research funding
Batten Disease Registry NYS Institute for Basic Research 1050 Forest Hill Road Staten Island, New York 10314 718-494-5201 fax 800-952-9628 battenkw@aol.com		Collects clinical, pathological and genetic data for research; provides expert diagnostic review of atypical cases
JNCL Research Fund P.O. Box 766 Mundelein, Illinois 60060	www.jnclresearch.org	Fund raising for research
Office of Communications and Public Liason National Institutes of Health Neurological Institute P.O. Box 5801 Bethesda, Maryland 301-496-5751 800-352-9424	www.ninds.nih.gov/ patients/disorder/ batten/battenfs.htm	Information about the diseases; support group contracts
University College London Medical School	www.ucl.ac.uk/ncl	Mutations

Foreknowledge of an inevitable illness when there is no therapy may rob the individual of a period of normalcy. However, testing in infancy or early childhood may limit the trauma of carrier identification, compared to learning of carrier status when the child is of marriageable age.

The incidence of the various NCLs varies from a high of about 1:20,000 for INCL in Finland (Uvebrandt and Hagberg, 1997) to less than 1:100,000. Since these are recessive conditions, at the low end the carrier frequency is about 1:71 and at the high end it is less than 1:158. For a non-Finnish unaffected sibling the a priori risk to have an affected child is $(2/3) \times (1/158) \times (1/4)$, about 1:1000, or 100 times greater than a person with no known risk.

In the past it was not possible to identify carriers reliably. Advances in molecular and biochemical testing now make it possible to identify carriers reliably in families where there is an affected child. It is more difficult to determine the carrier status of individuals in the general population in the United States by molecular means, because of the large number of mutations. In some countries, such as Finland, where a single molecular mutation accounts for the majority of cases, population screening to determine the carrier status of the general population is feasible. Carrier testing for late-infantile NCL in unrelated spouses is possible by biochemical analysis and finding about half of the expected enzyme activity (Junaid et al., 1999).

VII. REPRODUCTIVE OPTIONS

As with other recessive conditions, there are several reproductive options that need to be explained to families by the genetic counselor. These are listed in Table 8.2. Here we will limit discussion to prenatal diagnosis for the NCLs.

The earliest prenatal diagnosis attempts relied on identifying storage bodies by electron microscopy (MacLeod et al., 1984, 1985, 1988, Conradi et al., 1989; Kohlschutter et al., 1989; Rapola et al., 1988, 1990, 1993). However, these studies are technically challenging, and families cannot be fully reassured that the results are reliable (Lake, 1993; Chow et al., 1993; Goebel, 1995).

Using linkage, Järvelä et al. (1991a), were able to perform prenatal diagnosis for infantile NCL. Van Diggelin et al. (1999) showed retrospectively that palmitoyl-protein thioesterase (PPT) could be analyzed in CVS or AF. The combination of PPT activity and CLN1 mutation analysis in CVS identified a fetus affected with infantile NCL (de Vries et al., 1999).

The feasibility of prenatal testing for late-infantile NCL enzymatically was demonstrated by Junaid et al. (1999). Using both molecular mutation and biochemical analyses, Berry-Kravis (2000) excluded CLN2 in an at-risk pregnancy.

Table 8.2. Reproductive Options for Families with Childhood-Onset (Autosomal Recessive) Neuronal Ceroid Lipofuscinoses

- Have no additional children
- Accept 1:4 risk
- Adoption
- Prenatal diagnosis via chorionic villus sampling or amniocentesis
- Artificial insemination by donor
- In vitro fertilization and preimplantation testing of embryo

Combining electron microscopy and informative restriction length polymorphisms (RFLPs) flanking the CLN3 locus, Uvebrant et al. (1993) successfully excluded juvenile NCL with >99% confidence in CVS from an at-risk pregnancy. In a pregnancy in which the electron microscopy findings were suggestive of NCL, linkage disequilibrium of intragenic markers in CLN3 allowed for successful molecular diagnosis of juvenile NCL in the fetus of a Finnish woman. This diagnosis was further confirmed by identifying homozygosity of the common 1-kb mutation in the affected sibling and the affected fetus (Munroe et al., 1996). Tashner et al. (1997) developed an allele-specific PCR test to detect the major CLN3 deletion that was suitable for prenatal diagnosis and carrier detection.

Prenatal testing is not available for CLN4. Rapola et al. (1999), reported prenatal diagnosis of CLN5, the Finnish late-infantile variant. Electron microscopy of a CVS sample did not show inclusions, but the fetus was found to have the maternal 2-bp deletion and the paternal haplotype found in the affected proband. Following termination, the fetus was found to have inclusions consistent with variant late-infantile NCL.

There are 26 known mutations in CLN1, 31 in CLN2, 24 in CLN3, and 3 in CLN5 (www.ucl.ac.uk/ncl). For successful molecular prenatal diagnosis the mutations in the affected proband must be identified, preferably before a pregnancy. If the proband is not available, the DNA from the parents should be studied to identify each mutation. However, if it is possible to identify only one of the proband's mutations, prenatal diagnosis could be limited to exclusion. That is, if the fetus inherits the known mutation, there is a 50/50 chance that the fetus is affected. On the other hand, if the fetus does not inherit the known mutation, then the fetus could be only a carrier or normal.

The feasibility of preimplantation diagnosis for INCL was reported by Syvänen et al. (1997). They developed a solid-phase minisequencing test with whole-genome amplification suitable for reliable identification of the common Finnish infantile NCL mutation in single blastomeres.

VIII. CONCLUSION

Genetic counseling for the NCLs is complex. The diagnosis must be carefully confirmed and the specific form of NCL identified through biochemical or molecular genetic testing. The clinical course and prognosis need to be addressed. The risk of recurrence, including carrier testing issues and reproductive options, must be explained. Through all of this discussion, the counselor must be able to understand the psychosocial issues that are affecting the family and provide emotional support.

Acknowledgments

The author would like to thank Jacquelyn Krogh, M. S., Nancy Zellers, M. S., and Sarah Nolin, Ph.D., for their critical review of the manuscript.

References

Berry-Kravis, E., Sleat, D. E., Sohar, I., Meyer, P., Donnelly, R., and Lobel, P. (2000). Prenatal testing for late infantile neuronal ceroid lipofuscinosis. *Ann. Neurol.* **47,** 243–256.

Chow, C., Borg, J., Billson, V. R., and Lake, B. D. (1993). Fetal tissue involvement in the late infantile type of neuronal ceroid lipofuscinosis. *Prenat. Diagn.* **13,** 833–841.

Conradi, N. G., Uvebrant, P., Hökegård, L. H., Wahlström, H., and Mellqvist, L. (1989). First-trimester diagnosis of juvenile neuronal ceroid lipofuscinosis by demonstration of fingerprint inclusions in chorionic villi. *Prenat. Diagn.*, **9,** 283–287.

de Vries, B. B. A., Kleijer, W. J., Keulemans, J. L. M., Voznyi, Y. V., Franken, P. F., Eurlings, M. C. M., Galjaard, R. J., Losekoot, M., Catsman-Berrevoets, C. E., Breuning, M. H., Taschner, P. E. M., and van Diggelen, O. P. (1999). First-trimester diagnosis of infantile neuronal ceroid lipofuscinosis (INCL) using PPT enzyme assay and CLN1 mutation analysis. *Prenat. Diagn.* **19,** 559–562.

Fanos, J. H. (1997). Developmental tasks of childhood and adolescence: Implications for Genetic Testing. *Am. J. Med. Genet.* **71,** 22–28.

Gardiner, M., Sandford, A., Deadman, M., Poulton, J., Cookson, W., Reeders, S., Jokiaho, I., Peltonen, L., Eiberg, H., and Julier, C. (1990). Batten disease (Spielmeyer-Vogt disease, juvenile onset neuronal ceroid-lipofuscinosis) gene (CLN3) maps to human chromosome 16. *Genomics* **8,** 387–390.

Goebel, J. H., Vesa, J., Reitter, B., Goecke, T. O., Schneider-Rätzke, B., and Merz, E. (1995). Prenatal diagnosis of infantile neuronal ceroid-lipofuscinosis: A combined electron microscopic and molecular genetic approach. *Brain Dev.* **17,** 83–85.

Järvelä, I., Santavuori, P., Vesa, J., Rapola, J., Palotie, A., and Peltonen, L. (1991a). Infantile form of neuronal ceroid lipofuscinosis (CLN1) maps to the short arm of chromosome 1. *Genomics* **9,** 170–173.

Järvelä, I., Rapola, J., Peltonen, L., Puhakka, L., Vesa, J., and Ammälä, P., et al. (1991b). DNA-based prenatal diagnosis of the infantile form of neuronal ceroid lipofuscinosis INCL, CLN1). *Prenat. Diagn.* **11,** 323–328.

Junaid, M. A., Brooks, S. S., Wisniewski, K. E., and Pullarkat, R. K. (1999). A novel assay for lysosomal pepstatin-insensitive proteinase and its application for the diagnosis of late-infantile neuronal ceroid lipofuscinosis. *Clin. Chim. Acta* **281,** 169–176.

Kohlschütter, A., Rauskolb, R., Goebel, H. H., Anton, L. I., Albrecht, R., and Klein, H. (1989). Probable exclusion of juvenile neuronal ceroid lipofuscinosis in a fetus at risk: An interim report. *Prenat. Diagn.* **9,** 289–292.

Lake, B. D. (1993). Morphological approaches to the prenatal diagnosis of late-infantile and juvenile Batten disease. *J. Inher. Metab. Dis.* **16,** 345–348.

Macleod, P., Dolman, C., Nickel, R., Chang, E., Zonana, J., and Silvey, K. (1984). Prenatal diagnosis of neuronal ceroid lipofuscinosis. *N. Engl. J. Med.* **310,** 595.

MacLeod, P. M., Dolman, C. L., Nickel, R. E., Chang, E., Nag, S., Zonana, J., and Silvey, K. (1985). Prenatal diagnosis of neuronal ceroid-lipofuscinosies. *Am. J. Med. Genet.* **22,** 781–789.

MacLeod, P. M., Nag, S., and Berry, C. (1988). Ultrastructural studies as a method of prenatal diagnosis of neuronalceroid-lipofuscinosis. *Am. J. Med. Genet.* **Suppl. 5,** 93–97.

Munroe, P. B., Rapola, J., Mitchison, H. M., Mole, S. E., Gardiner, R. M., and Järvelä, I. (1999). Prenatal diagnosis of Batten's disease. *Lancet* **347,** 1014–1015.

Mole, S. E. (1999). Batten's disease: Eight genes and still counting? *Lancet* **354**, 443–445.
Munroe, P. B., Rapola, J., Mitchison, M., Mustonen, A., Mole, S. E., Gardiner, M., and Järevelä, I. (1996). Prenatal diagnosis of Batten's disease. *Lancet* **347**, 1014–1015.
Rapola, J., Santavuori, P., and Heiskala, H. (1988). Placental pathology and prenatal diagnosis of infantile neuronal ceroid-lipofuscinosis. *Am. J. Med. Genet.* **5**, 99–103.
Rapola, J., Salonen, R., Ammälä, P., and Santavuori, P. (1990). Prenatal diagnosis of the infantile type of neuronal ceroid lipofuscinosis by electron microscopic investigation of human chorionic villi. *Prenat. Diagn.* **10**, 553–559.
Rapola, J., Lähdetie, J., Isosomppi, J., Helminen, P., Penttinen, M., and Järvelä, I. (1999). Prenatal diagnosis of variant late infantile neuronal ceroid lipofuscinosis (vLINCL$_{Finnish}$; CLN5). *Prenat. Diagn.* **19**, 685–688.
Savukoski, M., Klockars, T., Holmberg, V., Santavuori, P., Lander, E. S., and Peltonen, L. (1998). CLN5, a novel gene encoding a putative transmembrane protein mutated in Finnish variant late infantile neuronal ceroid lipofuscinosis. *Nat. Genet.* **19**, 286–288.
Sharp, J. D., Wheeler, R. B., Lake, B. D., Savukoski, M., Järvelä, I. E., Peltonen, L., Gardiner, R. M., and Williams, R. E. (1997). Loci for classical and a variant late infantile neuronal ceroid lipofuscinosis map to chromosomes 11p15 and 15q21-23. *Hum. Mol. Genet.* **6**, 591–595.
Syvänen, A. C., Järvelä, I., Paunio, T., and Vesa, J. (1997). DNA diagnosis and identification of carriers of infantile and juvenile neuronal ceroid lipofuscinnosis. *Neuropediatrics* **28**, 63–66.
Tahvanainen, E.;, Ranta, S., Hirvasniemi, A., Karila, E., Leisti, J., Sistonen, P., Weissenbach, J., Lehesjoki, A.-E., and de la Chapelle, A. (1994). The gene for a recessively inherited human childhood progressive epilepsy with mental retardation maps to the distal short arm of chromosome 8. *Proc. Natl. Acad. Sci. (USA)* **91**, 7267–7270.
Tashner, P. E., deVos, N., and Breuning, M. H. (1997). Rapid detection of the major deletion in the Batten disease gene CLN3 by allele specific PCR. *J. Med. Genet.* **34**, 955–956.
Uvebrant, P., Björck, E., Conradi, N., Hökegård, K-H., Martinsson, T., and Wahlström, J. (1993). Successful DNA-based prenatal exclusion of juvenile neuronal ceroid lipofuscinosis. *Prenat. Diagn.* **13**, 651–657.
Van Diggelen, O. P., Keulemans, J. L., Winchester, B., Hofman, I. L., Vanhanen, S. L., Santavuori, P., and Voznyi, Y. V. (1999). A rapid fluorogenic palmitoyl-protein thioesterase assay: pre- and postnatal diagnosis of INCL. *Mol. Genet. Metab.* **66**, 240–244.
Wertz, D. C., Fanos, J. H., and Reilly, P. R. (1994). Genetic testing for children and adolescents: Who decides? *J. Am. Med. Assoc.*, **272**, 875–881.

9 Neurotrophic Factors as Potential Therapeutic Agents in Neuronal Ceroid Lipofuscinosis

Jonathan D. Cooper* and William C. Mobley
Department of Neurology and Neurological Sciences,
and the Program in Neuroscience
Stanford University School of Medicine
Stanford, California 94305

I. Introduction
II. Mouse Models of NCLs
III. Characterization of the CNS of Mouse Models of NCLs
IV. Neurotrophic Factors as Potential Therapeutic Agents in Neurodegenerative Disorders?—The Neurotrophic Factor Hypothesis
V. NTF Expression and Actions beyond the "Neurotrophic Factor Hypothesis"
VI. Failure of NTF Signaling—A Cause of Neuronal Dysfunction and Degeneration?
VII. Treatment with IGF-1—Implications for the Treatment of NCLs
VIII. Toward Clinical Trials of NTFs
References

*Present address and address for correspondence: Department of Neuropathology, Institute of Psychiatry, King's College London, De Crespigny Park, London, SE5 8AF, United Kingdom. E-mail: j.cooper@iop.kcl.ac.uk

I. INTRODUCTION

The neuronal ceroid lipofuscinoses (NCLs) are progressive, fatal neurodegenerative disorders of unknown pathogenesis with onset ranging from infancy to adulthood. Childhood forms are manifested by blindness, seizures, and dementia (Dyken, 1988; Goebel, 1995). Collectively these disorders represent the most common inherited neurodegenerative storage disorder of childhood, with an incidence of 1 in 12,500 live births (Goebel, 1995). The NCLs have traditionally been divided into four main types (Dyken, 1988; Goebel, 1995), although variant forms are reported (Dyken, 1988; Santavuori et al., 1991; Dyken and Wisniewski, 1995). In recent studies, genes have been discovered for the infantile form (CLN1) (Haltia-Santavuori disease) (Vesa et al., 1995), late-infantile form (CLN2) (Jansky-Bielschkowsky disease) (Sleat et al., 1997), juvenile form (CLN3) (Batten disease or Spielmeyer-Sjogren disease) (International Batten Disease Consortium, 1995), Finnish variant of the late-infantile form (CLN5) (Savukoski et al., 1998) and most recently, progressive epilepsy with mental retardation that was recently recognized as a new NCL subtype (CLN8) (Ranta et al., 1999). The gene for Kufs disease (CLN4), an adult-onset form of this disorder, awaits definition.

The characteristic feature of NCL pathology is the lysosomal accumulation of autofluorescent proteolipid in the brain and other tissues (Koenig, 1964; Haltia et al., 1973; Dyken, 1988; Goebel, 1995). Ultrastructurally, these electron-dense accumulations exhibit granular, curvilinear or fingerprint-like appearances, which are characteristic for each form of NCL (Santavuori, 1988; Goebel, 1995, 1997). Biochemical studies of these deposits in CLN2, CLN3 and CLN4 have shown that the major protein component is subunit c of the mitochondrial ATPase (Hall et al., 1991; Kominami et al., 1992; Palmer et al., 1992). In CLN1, deposits are composed largely of saposins A and D (Tyynelä et al., 1993). The molecular mechanisms by which mutations in the CLN genes lead to pathophysiology are, as yet, unidentified.

II. MOUSE MODELS OF NCLs

The development of an animal model that recapitulates the clinical and pathological features of NCLs represents an initial step toward discovering underlying disease mechanisms and testing potential treatment strategies. Although NCLs occur naturally in a range of animal species (Goebbel et al., 1999; Jolly, 1995) there is no animal model in which the disease is caused by mutations in CLN3. As such, it was necessary to utilize targeted disruption of the mouse Cln3 gene in embryonic stem cells to generate mice with a Cln3 null allele, a goal that has now been achieved by two different laboratories (Katz et al., 1999; Mitchison et al., 1999).

Initial characterization of both these strains of Cln3 −/− mice show progressive accumulation of autofluorescent material with the staining and ultrastructural characteristics of the material stored in Batten disease patients (Katz et al., 1999; Mitchison et al., 1999). More detailed analysis of one of these mouse lines has revealed that Cln2 protease activity in the brain was significantly elevated, implying that a correlation with human disease also exists at the biochemical level (Mitchison et al., 1999).

Two naturally occurring autosomal mutant mice, mnd and nclf, have also been shown to model late-infantile variant forms of the disorder. The mnd mouse was first characterized as exhibiting adult-onset upper and lower motor neuron degeneration (Messer and Flaherty, 1986; Messer et al., 1987), but closer examination revealed pathology similar to that seen in NCLs (Bronson et al., 1993; Mazurkiewicz et al., 1993; Pardo et al., 1994). The mice exhibit progressive retinopathy leading to blindness (Messer et al., 1993) and early, widespread accumulation of subunit c and autofluorescent lipopigment in many tissues (Bronson et al., 1993; Messer and Plummer, 1993; Pardo et al., 1994). Most recently, the mnd mouse has been shown to represent a naturally occurring model of cln8, a late-infantile variant form of NCL (Ranta et al., 1999). The mouse cln8 gene encodes a putative transmembrane protein, and in mnd bears a 1-bp insertion, which results in a frameshift and a truncated protein (Ranta et al., 1999). Another spontaneous mutant mouse, nclf, has also recently been identified which exhibits a grossly similar phenotype to the mnd/mnd mouse (Bronson et al., 1998). The nclf gene has been mapped to mouse chromosome 9 (Bronson et al., 1998), in a region syntenic to human chromosome 15q21 near the mapped locus for CLN6, another late-infantile variant form of NCL (Sharp et al., 1997).

III. CHARACTERIZATION OF THE CNS OF MOUSE MODELS OF NCLs

To determine which therapeutic strategies may be beneficial in reversing pathological changes and restoring neuronal function, it is essential to have a detailed understanding of which neuronal populations are affected in the NCLs and how their normal function is compromised. We have utilized the autofluorescent properties of accumulated lipopigment to reveal which neuronal populations show NCL-like pathological changes and examined the onset and progression of pathological changes in the CNS of cln3 null mutant mice, and the mnd and nclf mouse models of variant late-infantile NCL. In all three mutant mice we found a pronounced, early accumulation of autofluorescent lipopigment in subpopulations of GABAergic interneurons in the cortex and hippocampus (Cooper et al., 1999a; Mitchison et al., 1999, Lam et al., 1999). To determine the identity of those neurons that contained particularly dense accumulation of lipopigment, confocal microscopy was used to confirm the co-localization of lipopigment and immunofluorescence staining for several different markers of GABAergic phenotype, including the calcium-binding proteins parvalbumin (PV) and calbindin (Cb) and the neuropeptide somatostatin (SOM), which are all co-localized with GABA in many distinct subpopulations of GABAergic neurons (Freund and Buzsáki, 1996).

To ask whether lipopigment accumulation marked neurons whose phenotype is affected during disease progression, we examined populations of PV-positive neurons in the entorhinal cortex and subpopulations of PV-, Cb-, SOM-, and GAD-positive neurons in the hippocampal formation. In all three mutant mice we detected at least partial loss of staining for some, but not all, of these phenotypic markers, and pronounced hypertrophy of remaining detectable interneurons (Cooper et al., 1999a; Mitchison et al., 1999; Lam et al., 1999). Evidence from the spontaneous mutants *mnd* and *nclf* highlight the progressive nature of these morphological alterations, with more significant effects on both interneuronal number and size evident in both mutant strains. The extent of these changes differed markedly in these two mutants, with *mnd* showing more profound changes in the number of different subpopulations of hippocampal interneurons than *nclf* mice at 9 months of age (Cooper et al., 1999a; Lam et al., 1999).

Though the cellular changes in *Cln3* −/− mice were less severe than in *mnd* and *nclf* mice, we did observe significant loss of PV-positive neurons in the entorhinal cortex that is also reflected in the hippocampus, and a strong trend toward reduced numbers of SOM-positive neurons in the hippocampus (Mitchison et al., 1999). It is not yet clear whether the apparent loss of interneurons is due to cell death or the downregulation of normally expressed phenotypic markers (Cooper et al., 1999a). There was also significant hypertrophy of many subpopulations of interneurons, indicating dysregulation of cell volume (Mitchison et al., 1999).

Distinct effects of *Cln3* disruption on different interneuronal populations raise the possibility that different mechanisms may be operative. Since there is progressive cortical and cerebellar atrophy and selective loss of neurons in Batten disease, it will be important to examine the CNS of aged *Cln3* −/− mice for evidence of the progressive development of more significant pathological changes. Since these three mouse models of NCLs were generated in different background strains, it is difficult to make direct comparisons between *Cln3* −/− mice and the spontaneous mutants (*mnd* and *nclf*). However, it is apparent that GABA-ergic neurons are affected in each of these three mutant mice, albeit with different degrees of involvement in particular populations of interneurons in each mouse model. Indeed, it will be very informative to extend these studies to mouse models of other forms of the NCLs as they become available.

Inhibitory interneurons in the hippocampus and cortex exert a powerful influence on excitatory transmission in these brain regions (Freund and Buzsaki, 1996; Singer, 1996). It will be important to examine whether morphological abnormalities in any of these three mutant mice are reflected in compromised neuronal function. This question is particularly pertinent when considering the functional status of interneurons that exhibit profoundly abnormal morphology, which may be expected to be severely compromised in their ability to function normally. Perhaps the most clinically relevant questions to ask are whether such structural changes are accompanied by alterations in excitatory transmission or thresholds for seizure activation. We have already begun to investigate these issues in the *mnd* mouse, and have recently reported electrophysiological findings that suggest that functional inhibition in the dentate gyrus may be compromised in *mnd* mice that demonstrated significantly increased amplitude ratios and the presence of multiple population spikes (Cooper *et al.*, 1999b). Although these studies have already started *in vivo*, there is great scope for the extension of electrophysiological studies *in vitro*. This approach is particularly attractive since hippocampal slice methodology is very well established and lends itself to the screening of potential therapeutic and pharmacological manipulations that may directly address underlying pathophysiological mechanisms.

Our findings demonstrate that in all three mutant mice, progressively fewer GABA-ergic interneurons in both hippocampus and cortex can be detected by staining for markers that are normally present. The possibility arises that the reduction in number of immunohistochemically detected interneurons in these mice does not signify death, but instead marks downregulation of these markers to the extent that they are no longer detectable. Indeed, we found that many interneurons retained expression of GAD in the hippocampal formation of aged *mnd* mice, suggesting that at least some populations of these neurons remain alive despite the absence of other normally expressed phenotypic markers (Cooper *et al.*, 1999a). This raised the intriguing possibility that these populations of "phenotypically silent" interneurons might persist in a form that is amenable to therapeutic intervention.

IV. NEUROTROPHIC FACTORS AS POTENTIAL THERAPEUTIC AGENTS IN NEURODEGENERATIVE DISORDERS?—THE NEUROTROPHIC FACTOR HYPOTHESIS

Neurotrophic factors (NTFs) are polypeptides that support the growth, differentiation, and survival of neurons (Longo *et al.*, 1993; Bothwell, 1995; Yuen *et al.*, 1996). A number of NTFs have been discovered and their actions in the peripheral and central nervous system characterized in some detail (summarized in Table 9.1). The neurotrophins are an important family of this class of molecules. Nerve growth factor (NGF) is the prototypic member of the neurotrophin family of growth factors, the first-discovered and best-characterized NTF. Indeed, the discovery of NGF was instrumental to the formulation of the "neurotrophic factor hypothesis." This postulates that once a developing neuron has grown its process into

Table 9.1. Partial List of Neurotrophic Factors with Actions on Motor and Peripheral Nervous System (PNS) Neurons

Family	Member	Signaling receptor	Responsive neurons[a,b,c]
Neurotrophin	Nerve growth factor (NGF)	TrkA, p75[NTR]	PNS: 1, 2 (small sensory neurons), 3; CNS: 6, 7, 17, 19
	Brain-derived neurotrophic factor (BDNF)	TrkB, p75[NTR]	PNS: 1, 2; CNS: 5, 6, 8, 9, 11, 12, 18 20, 22, 25
	Neurotrophin-3 (NT-3)	TrkC, less so TrkA and TrkB p75[NTR]	PNS: 1, 2 (large sensory neurons), 3; CNS: 5, 6, 9, 11, 18, 20, 22, 24, 25
	Neurotrophin-4/5 (NT-4/5)	TrkB, p75[NTR]	PNS: 1–3; CNS: 5, 6, 9, 11 12, 18, 20, 24, 25
Neuropoietic cytokine	Ciliary neurotrophic factor (CNTF) eukemia inhibitory (LIF or CDF/LIF)	CNTF receptor (CNTFRα, gp130, LIFRβ subunits) IF receptor (gp130, LIFRβ subunits)	PNS: 1, 3, 4; CNS: 5, 6, 11, 13, 14, 18–20, 24 PNS: 1–3, CNS: 5
Insulin-like growth factor (IGF)	IGF-1	Type I IGF receptor (IGFIR) less so insulin receptor (IR)	PNS: 1, 3; CNS: 5, 6, 11, 14, 15, 25, 26
	IGF-2	IGFIR and less so IR	PNS: 1, 3; CNS: 5, 6
Transforming growth factor-β (TGF-β) family	TGF-β1, TGF-β2, TGF-β3	TGF-β types I, II, and III receptors	PNS: 1; CNS: 5, 11, 14, 18
	Glial cell line-derived neurotrophic factor (GDNF)	ret, GDNF family receptor (GFR) α-1 and α-2	PNS: 1–5, CNS: 5, 11, 13, 22

	Neurturin (NTN)	ret, GFR α-2 and α-1	PNS: 1–5, CNS: 5, 11, and possibly 13, 24
	Persephin (PSP)	ret, GFR α-4	CNS: 5, 11
	Artemin (ARTN)	ret, GFR α-3 and α-1	PNS: 1–3; CNS: 11
Fibroblast growth factor (FGF)	Acidic FGF (aFGF or FGF-1)	FGF receptors 1–4 (FGFRs 1–4)	PNS: 1, 3, 4; CNS: 5, 14, 16, 18–20, 23, 25
	Basic FGF (bFGF or FGF-2)	FGFRs 1–3	PNS: 1, 3, 4; CNS: 5, 11, 14, 16, 18–21, 25
	FGF-5	FGFR-1, FGFR-2	PNS: None known; CNS: 5
Other growth factors	Transforming growth factor alpha (TGF-α)	Epidermal growth factor receptor	PNS: 1; CNS: 11
	Platelet-derived growth factor (PDGF: AA, AB BB isoforms)	PDGF α- and β-receptors	PNS: None known; CNS: 5, 10, 11, 20
	Stem cell factor	c-kit	PNS: 1; CNS: 15 (possibly)

^aPNS: (1) DRG sensory neurons; (2) cranial sensory neurons; (3) sympathetic neurons; (4) parasympathetic neurons; (5) enteric neurons.

^bCNS—specific neuronal populations: (5) motor neurons; (6) basal forebrain cholinergic neurons; (7) striatal cholinergic neurons; (8) basal forebrain γ-aminobutyric (GABA)-ergic neurons; (9) mesencephalic GABA-ergic neurons; (10) cerebellar GABA-ergic neurons; (11) substantia nigra dopaminergic neurons; (12) cerebellar granule cells; (13) Purkinje cells.

^cCNS—neurons of the: (14) cortex, (15) cerebellum, (16) spinal cord, (17) brainstem, (18) hippocampus, (19) thalamus, (20) striatum, (21) septum, (22) trigeminal mesencephalic nucleus, (23) subiculum, (24) locus coeruleus, (25) retinal ganglion, (26) olfactory bulb.

its target, it competes with other developing neurons of the same type for a limited supply of a NTF provided by the target. Successful competitors survive; unsuccessful ones die. Under this hypothesis, a NTF should be diffusible and act on specific cell surface receptors localized to the processes of innervating neurons. It also predicts that a NTF(s) will be present at a concentration lower than that necessary to maintain the viability of all innervating neurons. In our view, reasonable extensions of the hypothesis are that: (1) the target would produce one or only a limited number of NTFs for each innervating population; (2) correspondingly, there could be a specific NTF for each neuronal population—literally thousands of such factors would exist to act on the many thousands of different neuronal populations in the nervous system; (3) neurons would be the only cells to respond to a NTF; and (4) as for survival, differentiation would also be regulated by limiting concentrations of NTFs. Although important aspects of NTFs and their signaling biology were accurately predicted by the "neurotrophic factor hypothesis," there is considerably more complexity and diversity than was initially anticipated.

V. NTF EXPRESSION AND ACTIONS BEYOND THE "NEUROTROPHIC FACTOR HYPOTHESIS"

Studies on NTFs have considerably broadened our knowledge about their actions. Neurotrophins and other NTFs act by binding to specific cell surface receptors. Activated receptors initiate intracellular signaling events, leading to induction of gene expression and changes in neuronal morphology and function. Many investigations have been directed at characterizing these signaling events, not only for the neurotrophins but for many other NTFs including ciliary neurotrophic factor (CNTF), the insulin-like growth factors (IGFs), the fibroblast growth factors (FGFs), and numbers of the glial cell line-derived nerve factor (GDNF) family (Yuen et al., 1996; Baloh et al., 1998) (Table 9.1). There is much that was not predicted by the original formulation of the "neurotrophic factor hypothesis." For example, it is now evident from data for the expression and actions of neurotrophins that, in addition to target-derived delivery, autocrine and non-target-derived paracrine modes of presentation are likely to be important (Schecterson and Bothwell, 1992). Another departure from the original "neurotrophic factor hypothesis" is evidence that an individual population of neurons can respond to many different NTFs. As such, it is possible that several factors acting together determine whether particular neuronal populations live or die. An interesting correlate is data showing that a single factor can act on many different neuronal populations. Thus the expectation that thousands of NTFs would be required to direct the development of an equally large number of neuronal populations may be incorrect. Instead, precise combinations of a limited number of factors may be important. It is also apparent that certain NTFs act on non-neuronal cells

(Yuen et al., 1996; Lindsay et al., 1996). Indeed, many NTFs were first discovered through their actions on such cells.

The "neurotrophic factor hypothesis" created the expectation that specific NTFs would act through specific receptors. While this is the case, there is considerably more diversity than expected. Recently, it has become apparent that the actions of NTFs are not restricted to developing postmitotic neurons. Rather, these factors can also act on dividing neuroblasts as well as on mature neurons that have established stable synaptic contacts. NTFs are now known to play a role in the functional modification of synaptic connections within the central and peripheral nervous systems (Berninger and Poo, 1996; Bonhoeffer, 1996; Lo, 1995; Stoop and Poo, 1996). NTF actions at central nervous system (CNS) synapses are particularly interesting. NTF secretion is increased by neuronal depolarization or synaptic activation (Blöchl and Theonen, 1995; Gu et al., 1994; Wang and Poo, 1997), and neurotrophic factor expression within the CNS is regulated by electrical activity (Ballarín et al., 1991; Gall and Isackson, 1989; Gwag and Springer, 1993; Kesslak et al., 1998; Knipper et al., 1994; Patterson et al., 1992; Theonen et al., 1991; Wetmore et al., 1994; Zafra et al., 1991). Moreover, there is widespread cortical and hippocampal expression of neurotrophins and their cognate receptors (Barbacid, 1994; Bothwell, 1995; Phillips et al., 1990; Hofer et al., 1990; Wetmore et al., 1990). These findings suggest that neurotrophins may play a role in synaptic plasticity within the CNS. These new data give exciting hints regarding a role for neurotrophins modifying the structure and function of synaptic connections, a role not anticipated in the "neurotrophic factor hypothesis."

VI. FAILURE OF NTF SIGNALING—A CAUSE OF NEURONAL DYSFUNCTION AND DEGENERATION?

Studies of the cell biology of NTF signaling are creating new possibilities for exploring the pathogenesis of neurodegeneration. Our laboratory has long been interested in the degeneration of basal forebrain cholinergic neurons (BFCNs) in Alzheimer's disease and elderly Down syndrome patients, and we have entertained the hypothesis that failed NGF signaling contributes to this phenotype. In recent studies in a mouse model of Down syndrome, the partial-trisomy 65Dn mouse, we discovered that there is age-related atrophy and apparent loss of these neurons (Holtzman et al., 1996). Interestingly, in spite of normal levels of NGF in the hippocampal target of these cells, and apparently normal hippocampal innervation by cholinergic axons, NGF levels in the basal forebrain were decreased (J. D. Cooper and W. C. Mobley, unpublished observations). The signaling endosome hypothesis argues that retrograde transport of NGF marks the retrograde NGF signal (Grimes et al., 1996, 1997). If so, our findings would indicate that BFCNs in partial-trisomy 65Dn mice receive less NGF signaling from the target and are thereby trophically deprived. Significantly, in view of the important role for NGF in the growth and maintenance of BFCNs, failed NGF signaling could explain the dysfunction and death of these neurons. We have shown that delivering NGF to BFCN cell bodies reverses atrophy and the apparent loss of these cells (J. D. Cooper and W. C. Mobley, unpublished observations).

VII. TREATMENT WITH IGF-1—IMPLICATIONS FOR THE TREATMENT OF NCLs

Collectively, NTFs represent a group of molecules that exert a powerful influence on neuronal development, differentiation, and survival, in addition to regulating many aspects of normal neuronal biology. We were interested in whether treatment with an NTF might be capable of reversing pathology involving interneurons in *mnd* mice. Insulin-like growth factor-1 (IGF-1) was previously shown to have beneficial effects on detectable GABA-ergic interneuronal number and dendritic morphology in a canine tissue culture model of NCL (Dunn et al., 1994). Moreover, expression of IGF receptors is widespread throughout the CNS (Adamo et al., 1989) and, as such, treatment with IGF can be considered as targeting one specific trophic factor to many neuronal populations. Therefore, we compared the effect of 7 days of treatment by intracerebroventricular infusion with 2 μg/day IGF-1 or artificial cerebrospinal fluid vehicle in aged control and *mnd* animals on interneuronal number in the entorhinal cortex and hippocampal formation.

Remarkably, treatment with IGF-1 significantly increased the number of SOM-positive hilar interneurons compared with age-matched, vehicle-treated *mnd* mice (Cooper et al., 1999a). The effect of IGF-1 treatment on cross-sectional area was more robust than the effect on neuronal number, significantly reducing the size of SOM-positive hilar interneurons, PV-positive neurons in the stratum oriens, and PV-positive neurons of the dentate gyrus (Cooper et al., 1999a). In spite of this, we detected no obvious difference in IGF-1 treated animals in the density of lipopigment deposits in SOM- or PV-positive neuronal cell bodies or in the morphology of their dendrites.

It is remarkable that just 1 week of IGF-1 treatment was effective in partially restoring interneuronal number and reducing hypertrophy in some subhippocampal subregions (Cooper et al., 1999a). These findings are evidence that at least some "phenotypically silent" neurons are growth factor-responsive. Indeed IGF-1 treatment acted to reverse the neurodegenerative phenotype in responsive cells, suggesting that IGF-1 could have a role in therapy. However, IGF-1 treatment had no effect on the size or number of PV-positive interneurons in the entorhinal cortex, suggesting that other mechanisms may operate in this brain region, that degenerative changes had become irreversible before treatment was commenced, or that a longer course of treatment would be required to see an effect. It will be informative to test further the effect of IGF-1 and other neurotrophic factors, particularly at younger ages, to test whether a more robust effect in reversing the degenerative changes can be produced. Although our findings have direct implications for devising potential therapeutic strategies for the treatment of patients with NCLs, rigorous tests of candidate strategies in appropriate animal models will be required.

VIII. TOWARD CLINICAL TRIALS OF NTFs

Robust NTF effects on survival and differentiation of normal neurons have suggested that these factors may be used to enhance the survival and function of neurons threatened by neurological diseases. In recent years, animal models of human neurological disorders have been used to test the therapeutic potential of NTFs. In a line of investigation that was not predicted by the "neurotrophic factor hypothesis" but followed logically from it, NTFs have been shown to protect against neuronal dysfunction and death in several animal models of injury and neurological disease (Yuen et al., 1996). The emergence of NTFs as potential therapies for neurological disorders is exciting but not without certain problems, as the technicalities of appropriate and targeted delivery remain unsolved. However, systematic study may yet reveal the true potential of NTFs as therapeutic agents to treat neurological disorders including the NCLs.

Acknowledgments

J. D. Cooper and W. C. Mobley thank their colleagues, Jane Chua-Couzens, Andrew Feng, Dawn Lam, Pavel Belichenko, Ahmad Salehi, Josh Kilbridge, Janice Valletta, Alfredo Ramirez, Arthur Lee, David Holtzman, Hannah Mitchison, Nick Greene, Robert Nussbaum, Paul Buckmaster, and Anne Messer, for their contributions to this work, and Drs. Alison Barnwell and Pavel Belichenko for critical review of the manuscript. This work was supported by National Institutes of Health grant NS29110 (A. M.), The Remy foundation (J. D. C.), The Batten's Disease Support and Research Association, The Natalie Fund and the Children's Brain Diseases Foundation (W. C. M). IGF-1 was provided by Dr. Nicola Neff of Cephalon, Inc.

References

Adamo, M., Raizada, M. K., and LeRoith, D. (1989). Insulin and insulin-like growth factor receptors in the nervous system. *Mol. Neurobiol.* **3,** 71–100.

Ballarín, M., Ernfors, P., Lindefors, N., and Persson, H. (1991). Hippocampal damage and kainic acid injection induce a rapid increase in mRNA for BDNF and NGF in the rat brain. *Exp. Neurol.* **114,** 35–43.

Baloh, R. H., Tansey, M. G., Lampe, P. A., Fahrner, T. J., Enomoto, H., Simburger, K. S., Leitner, M. L., Araki, T., Johnson, E. M., and Milbrandt, J. (1998). Artemin, a novel member of the GDNF ligand family, supports peripheral and central neurons and signals through the GFRalpha3-RET receptor complex. *Neuron* **21,** 1291–1302.

Barbacid, M. (1994). The Trk family of neurotrophin receptors. *J. Neurobiol.* **25,** 1386–1403.

Berninger, B., and Poo, M. (1996). Fast actions of neurotrophic factors. *Curr. Opin. Neurobiol.* **6,** 324–330.

Blöchl, A., and Thoenen, H. (1995). Characterization of nerve growth factor (NGF) release from hippocampal neurons: Evidence for a constitutive and an unconventional sodium-dependent regulated pathway. *Eur. J. Neurosci.* **7,** 1220–1228.

Bonhoeffer, T. (1996). Neurotrophins and activity-dependent development of the neocortex. *Curr. Opin. Neurobiol.* **6,** 119–126.

Bothwell, M. (1995). Functional interactions of neurotrophins and neurotrophin receptors. *Annu. Rev. Neurosci.* **18,** 223–253.

Bronson, R. T., Lake, B. D., Cook, S., Taylor, S., and Davisson, M. T. (1993). Motor neuron degeneration of mice is a model of neuronal ceroid lipofuscinosis (Batten's disease). *Ann. Neurol.* **33,** 381–385.

Bronson, R. T., Donahue, L. R., Johnson, K. R., Tanner, A., Lane, P. W., and Faust, J. R. (1998). Neuronal ceroid lipofuscinosis (*nclf*), a new disorder of the Mouse linked to chromosome 9. *Am. J. Med. Genet.* **77,** 289–297.

Cooper, J. D., Messer, A., Feng, A. K., Chua-Couzens, J., and Mobley, W. C. (1999a). Apparent loss and hypertrophy of interneurons in a mouse model of neuronal ceroid lipofuscinosis: Evidence for partial response to insulin-like growth factor-1 treatment. *J. Neurosci.* **19,** 2556–2567.

Cooper, J. D., Mobley, W. C., and Buckmaster, P. S. (1999b). Compromised functional inhibition in the hippocampus of the *mnd* mouse model of neuronal ceroid lipofuscinosis. *Soc. Neurosci. Abstr.* **25,** 1593.

Dunn, W. A., Raizada, M. K., Vogt, E. S., and Brown, E. A. (1994). Growth factor-induced growth in primary neuronal cultures of dogs with neuronal ceroid lipofuscinosis. *Int. J. Dev. Neurosci.* **12,** 185–196.

Dyken, P. (1988). Reconsideration of the classification of the neuronal ceroid lipofuscinoses. *Am. J. Med. Genet.* **Suppl. 5,** 69–84.

Dyken, P., and Wisniewski, K. (1995). Classification of the neuronal ceroid-lipofuscinoses: Expansion of the atypical forms. *Am. J. Med. Genet.* **57,** 150–154.

Freund, T. F., and Buzsaki, G. (1996) Interneurons of the hippocampus. *Hippocampus* **6,** 347–470.

Gall, C. M., and Isackson, P. J. (1989). Limbic seizures increase neuronal production of messenger RNA for nerve growth factor. *Science* **245,** 758–761.

Goebel, H. H. (1995). The neuronal ceroid-lipofuscinoses. *J. Child. Neurol.* **10,** 424–437.

Goebel, H. H. (1997). Morphologic diagnosis in neuronal ceroid lipofuscinosis. *Neuropediatrics* **28,** 167–169.

Goebel, H. H., Mole, S. E., and Lake, B. D. (eds.) (1999). "The Neuronal Ceroid Lipofuscinoses (Batten Disease)," Biomedical and Health Research Vol. 33. IOS Press, Amsterdam, The Netherlands.

Grimes, M. L., Zhou, J., Beattie, E. C., Yuen, E. C., Hall, D. E., Valletta, J. S., Topp, K. S., LaVail, J. H., Bunnett, N. W., and Mobley, W. C. (1996). Endocytosis of activated TrkA: Evidence that nerve growth factor induces formation of signaling endosomes. *J. Neurosci.* **16,** 7950–7964.

Grimes, M. L., Beattie, E. C., and Mobley, W. C. (1997). A signaling organelle containing the nerve growth factor-activated receptor tyrosine kinase, TrkB. *Proc. Natl. Acad. Sci. (USA)* **94,** 990–914.

Gu, Q., Liu, Y., and Cynader, M. S. (1994). Nerve growth factor-induced ocular dominance plasticity in adut cat visual cortex. *Proc. Natl. Acad. Sci. (USA)* **91,** 8408–8412.

Gwag, B. J., and Springer, J. E. (1993). Activation of NMDA receptors increases brain-derived neurotrophic factor (BDNF) mRNA expression in the hippocampal formation. *Neuroreport* **5,** 125–128.

Hall, N. A., Lake, B. D., Dewji, N. N., and Patrick, AD (1991). Lysosomal storage of subunit c of mitochondrial ATP synthase in Batten's disease (ceroid-lipofuscinosis). *Biochem. J.* **275,** 269–272.

Haltia, M., Rapola, J., Santavuori, P., and Keranen, A. (1973). Infantile type of so-called neuronal ceroid-lipofuscinosis. 2. Morphological and biochemical studies. *J. Neurol. Sci.* **18,** 269–285.

Hofer, M., Pagliusi, S. R., Hohn, A., Leibrock, J., and Barde, Y. A. (1990). Regional distribution of brain-derived neurotrophic factor mRNA in the adult mouse brain. *EMBO J.* **9,** 2459–2464.

Holtzman, D. M., Santucci, D., Kilbridge, J., Chua-Couzens, J., Fontana, D. J., Daniels, S. E., Johnson, R. M., Chen, K., Sun, Y., Carlson, E., Alleva, E., Epstein, C. J., and Mobley, W. C. (1996). Developmental abnormalities and age-related neurodegneration in a mouse model of Down syndrome. *Proc. Natl. Acad. Sci. (USA)* **93,** 13333–13338.

International Batten Disease Consortium (1995). Isolation of a novel gene underlying Batten disease, *CLN3. Cell* **82,** 949–957.

Jolly, R. D. (1995). Comparative biology of the neuronal ceroid lipofuscinoses (NCL): An overview. *Am. J. Med. Genet.* **57,** 307–311.

Katz, M. L., Shibuya, H., Liu, P. C., Kaur, S., Gao, C. L., and Johnson, G. S. (1999). A mouse gene knockout model for juvenile ceroid-lipofuscinosis (Batten disease). *J. Neurosci. Res.* **57,** 551–556

Kesslak, J. P., So, V., Choi, J., Cotman, C. W., and Gomez-Pinilla, F. (1998). Learning upregulates brain-derived neurotrophic factor messenger ribonucleic acid: A mechanism to facilitate encoding and circuit maintenance? *Behav. Neurosci.* **112,** 1012–1019.

Koenig, H. (1964). Neuronal lipofuscin in disease. Its relation to lysosomes. *Trans. Am. Neurol. Assoc.* **89,** 212–213.

Kominami, E., Ezaki, J., Muno, D., Ishido, K., Ueno, T., and Wolfe, L. S. (1992). Specific storage of subunit c of mitochondrial ATP synthase in lysosomes of neuronal ceroid lipofuscinosis (Batten's disease). *J. Biochem.* **111,** 278–282.

Knipper, M., da Penha Berzaghi, M., Blöchl, A., Breer, H., Thoenen, H., and Lindholm, D. (1994). Positive feedback between acetylcholine and the neurotrophins nerve growth factor and brain-derived neurotrophic factor in the rat hippocampus. *Eur. J. Neurosci.* **6,** 668–671.

Lam, H. H. D., Mitchison, H. M., Greene, N. D. E., Nussbaum, R. L., Mobley, W. C., and Cooper, J. D. (1999). Pathologic involvement of interneurons in mouse models of neuronal ceroid lipofuscinosis. *Soc. Neurosci. Abstr.* **25,** 1593.

Lindsay, R. M., and Yancopoulos, G. D. (1996). GDNF in a bind with known orphan: Accessory implicated in new twist. *Neuron* **17,** 571–574.

Lo, DC. (1995). Neurotrophic factors and synaptic plasticity. *Neuron* **15,** 979–981.

Longo, F. M., Holtzman, D. M., Grimes, M. L., and Mobley, W. C. (1993). Nerve growth factor: Actions in the peripheral and central nervous systems. *In* "Neurotrophic Factors" (J. Fallon and S. Loughlin, eds.), pp. 209–256. Academic Press, New York.

Mazurkiewicz, J. E., Callahan, L. M., Swash, M., Martin, J. E., and Messer, A. (1993). Cytoplasmic inclusions in spinal neurons of the motor neuron degeneration (Mnd) mouse. *J. Neurol. Sci.* **116,** 59–66.

Messer, A., and Flaherty, L. (1986). Autosomal dominance in a late-onset motor neuron disease in the mouse. *J. Neurogenet.* **3,** 345–355.

Messer, A., and Plummer, J. (1993). Accumulating autofluorescent material as a marker for early changes in the spinal cord of the Mnd mouse. *Neuromusc. Disord.* **3,** 129–134.

Messer, A., Strominger, N. L., and Mazurkiewicz, J. E. (1987). Histopathology of the late-onset motor neuron degeneration (Mnd) mutant in the mouse. *J. Neurogenet.* **4,** 201–213.

Messer, A., Plummer, J., Wong, V., and La Vail, M. M. (1993). Retinal degeneration in motor neuron degeneration (mnd) mutant mice. *Exp. Eye. Res.* **57,** 637–641.

Mitchison, H. M., Bernard, D. J., Greene, N. D., Cooper, J. D., Junaid, M. A., Pullarkat, R. K., de Vos, N., Breuning, M. H., Owens, J. W., Mobley, W. C., Gardiner, R. M., Lake, B. D., Taschner, P. E., and Nussbaum, R. L. (1999). Targeted disruption of the *Cln3* gene provides a mouse model for Batten disease. *Neurobiol. Dis.* **6,** 321–34.

Palmer, D. N., Fearnley, I. M., Walker, J. E., Hall, N. A., Lake, B. D., Wolfe, L. S., Haltia, M., Martinus, R. D., and Jolly, R. D. (1992). Mitochondrial ATP synthase subunit c storage in the ceroid-lipofuscinoses (Batten disease). *Am. J. Med. Genet.* **42,** 561–567.

Pardo, C. A., Rabin, B. A., Palmer, D. N., and Price, D. L. (1994). Accumulation of the adenosine triphosphate synthase subunit c in the *mnd* mutant mouse. *Am. J. Pathol.* **144,** 829–835.

Patterson, S. L., Grover, L. M., Schwartzkroin, P. A., and Bothwell, M. (1992). Neurotrophin expression in rat hippocampal slices: A stimulus paradigm inducing LTP in CA1 evokes increases in BDNF and NT-3 mRNAs. *Neuron* **9,** 1081–1088.

Phillips, H. S., Hains, J. M., Laramee, G. R., Rosenthal, A.,, and Winslow, J. W. (1990). Widespread expression of BDNF but not NT3 by target areas of basal forebrain cholinergic neurons. *Science* **250,** 290–294.

Ranta, S., Zhang, Y., Ross, B., Lonka, L., Takkunen, E., Messer, A., Sharp, J., Wheeler, R., Kusumi, K., Mole, S., Liu, W., Soares, M. B., Bonaldo, M. F., Hirvasniemi, A., de la Chapelle, A., Gilliam, T. C., and Lehesjoki, A. E. (1999). The neuronal ceroid lipofuscinoses in human EPMR and mnd mutant mice are associated with mutations in CLN8. *Nat. Genet.* **23,** 233–236.

Santavuori, P. (1988). Neuronal ceroid lipofuscinosis in childhood. *Brain Dev.* **10,** 80–83.

Santavuori, P., Rapola, J., Nuutila, A., Raininko, R., Lappi, M., Launes, J., Herva, R., and Saino, K. (1991). The spectrum of Jansky-Bielschowsky disease. *Neuropediatrics* **22,** 92–96.

Savukoski, M., Klockars, T., Holmberg, V., Santavuori, P., Lander, E. S., and Peltonen, L. (1998). CLN5, a novel gene encoding a putative transmembrane protein mutated in Finnish variant late infantile neuronal ceroid lipofuscinosis. *Nat. Genet.* **19,** 286–288.

Schecterson, L. C., and Bothwell, M. (1992). Novel roles for neurotrophins are suggested by BDNF and NT-3 mRNA expression in developing neurons. *Neuron* **9,** 449–463.

Sharp, J. D., Wheeler, R. B., Lake, B. D., Savukoski, M., Jarvela, J., Peltonen, L., Gardiner, R. M., and Williams, R. E. (1997). Loci for classical and a variant late infantile neuronal ceroid lipofuscinosis map to chromosomal 11p15 and 15q21-23. *Hum. Mol. Genet.* **6,** 591–596

Singer, W. (1996). Neurophysiology: The changing face of inhibition. *Curr. Biol.* **6,** 395–397.

Sleat, D., Donnelly, R. J., Lackland, H., Liu, C.-G., Sohar, I., Pullarkat, R. K., and Lobel, P. (1997). Association of mutations in a lysosomal protein with classical late-infantile neuronal ceroid lipofuscinosis. *Science* **277,** 1802–1805.

Stoop, R., and Poo, M. M. (1996). Synaptic modulation by neurotrophic factors. *Prog. Brain Res.* **109,** 359–364.

Thoenen, H., Zafra, F., Hengerer, B., and Lindholm, D. (1991). The synthesis of nerve growth factor and brain-derived neurotrophic factor in hippocampal and cortical neurons is regulated by specific transmitter systems. *Ann. N.Y. Acad. Sci.* **640,** 86–90.

Tyynelä, J., Palmer, D. N., Baumann, M., and Haltia, M. (1993). Storage of saposins A and D in infantile neuronal ceroid-lipofuscinosis. *FEBS Lett.* **330,** 8–12.

Vesa, J., Hellsten, E., Verkruyse, L. A., Camp, L. A., Rapola, J., Santavuori, P., Hofmann, S. L., and Peltonen, L. (1995). Mutations in the palmitoyl protein thioesterase gene causing infantile neuronal ceroid lipofuscinosis. *Nature* **376,** 584–587.

Wang, X. H. and, and Poo, M. M. (1997). Potentiation of developing synapses by postsynaptic release of neurotrophin-4. *Neuron* **19,** 925–935.

Wetmore, C., Ernfors, P., Persson, H., and Olson, L. (1990). Localization of brain-derived neurotrophic factor mRNA to neurons in the brain by in situ hybridization. *Exp. Neurol.* **109,** 141–152.

Wetmore, C., Olson, L. and, and Bean, A. J. (1994). Regulation of brain-derived neurotrophic factor (BDNF) expression and release from hippocampal neurons is mediated by non-NMDA type glutamate receptors. *J. Neurosci.* **14,** 1688–1700.

Yuen, E. C., Howe, C. L., Yiwen, L., Holtzman, D. M., and Mobley, W. C. (1996). Nerve growth factor and the neurotrophic factor hypothesis. *Brain Dev.* **18,** 362–368.

Zafra, F., Castrén, E., Thoenen, H., and Lindholm, D. (1991). Interplay between glutamate and gamma-aminobutyric acid transmitter systems in the physiological regulation of brain-derived neurotrophic factor and nerve growth factor synthesis in hippocampal neurons. *Proc. Natl. Acad. Sci. (USA)* **88,** 10037–10041.

10

Animal Models for the Ceroid Lipofuscinoses

Martin L. Katz[*]
University of Missouri School of Medicine
Mason Eye Institute
Columbia, Missouri 65212

Hisashi Shibuya
Department of Veterinary Medicine
Nihon University
Fujisawa, Japan 252-8510

Gary S. Johnson
Department of Veterinary Pathobiology
University of Missouri College of Veterinary Medicine
Columbia, Missouri 65211

I. The Need for Animal Models
II. The Human Disorders
III. Naturally Occurring Ceroid Lipofuscinosis in Animals as Models for the Human Disorders
   A. English Setter Dogs
   B. South Hampshire Sheep
   C. Tibetan Terrier Dogs
   D. *mnd* Mice
   E. *nclf* Mice
IV. Animal Models Created through Molecular Genetic Manipulation
   A. Mouse Gene Knockout Models of Juvenile NCL *(CLN3)*
   B. Mouse Gene Knockout Model of Late-Infantile NCL
V. Future Directions
   References

[*]Address for correspondence: E-mail: katzm@health.missouri.edu

I. THE NEED FOR ANIMAL MODELS

As with many human maladies, the search for effective treatments or cures for the various ceroid lipofuscinoses would be greatly facilitated if appropriate animal models were available for studies on the mechanisms underlying the disease pathologies and for screening potential therapeutic interventions. This is particularly true for relatively rare diseases for which it would be difficult to amass a sufficiently large pool of human subjects for controlled clinical intervention trials. Since the recognition of the human ceroid lipofuscinoses as distinct disease entities, a number of naturally occurring diseases with similar features have been described in animals (Bildfell et al., 1995; Bronson et al., 1993; Chang et al., 1994; Goebel et al., 1988; Jarplid and Haltia, 1993; Jolly, 1995; Jolly et al., 1992a, 1992b; Jolly and Palmer, 1995; Koppang, 1988, 1992; Palmer et al., 1997; Tayler and Farrow, 1992; Weissenbock and Rossel, 1997). Some of these animals have been used in research to elucidate the mechanisms involved in the human disease pathologies (Jolly et al., 1980, 1988; Koppang, 1988; Palmer et al., 1986a, 1986b, 1988, 1989, 1990). However, it was not until the relatively recent identification of the gene defects involved in the human disorders (International Batten Disease Consortium, 1995; Sleat et al., 1997; Vesa et al., 1995) that it was possible to evaluate whether the naturally occurring animal diseases are good models for the human counterparts. Identification of the genes involved in the human disorders also provided the information necessary for creating animal models through molecular genetic manipulation.

II. THE HUMAN DISORDERS

The hereditary neuronal ceroid lipofuscinoses (NCLs) are autosomal recessively inherited neurodegenerative diseases that have a number of clinical and pathological features in common (Boustany et al., 1988; Wisniewski et al., 1988). Despite their similarities, however, the NCLs are actually a group of distinct disorders that result from defects in different genes. The most common diseases that are classified as NCLs result from defects in the *CLN1*, *CLN2*, and *CLN3* genes associated with what are commonly referred to as infantile, late-infantile, and juvenile NCL, respectively. Defects in at least five other genes cause NCL-like disorders (Dyken and Wisniewski, 1995; Dyken, 1988; Klockars et al., 1996; Mole, 1999; Savukoski et al., 1998; Sharp et al., 1997). Given this genetic heterogeneity, one must ask whether a single animal model will be adequate to address questions of disease mechanisms and evaluate therapeutic interventions, or whether animal models will be needed to mimic each specific gene defect. To address the latter question, it will be necessary to review briefly what is known about each of the more common human NCLs.

The infantile disease (CLN1) is characterized by an early onset and rapid progression of clinical symptoms. As the name implies, the disorder becomes apparent in infancy, usually before the age of 1 year. Symptoms include prominent myoclonus, visual loss due to retinal atrophy, dementia, and motor deterioration (Wisniewski et al., 1988). Eventually, affected children become vegetative and may remain in this state for years before dying between the ages of 8 and 13 years (Boustany et al., 1988; Wisniewski et al., 1988). Lysosomal storage bodies with contents that have granular ultrastructural appearances accumulate in many tissues of affected individuals. Among the molecular constituents of the storage bodies are large amounts of sphinglipid activator proteins (SAPs) (Tyynela et al., 1993, 1995, 1997a). The defective gene in children with infantile NCL was identified as the palmitoyl-protein thioesterase 1 gene (*PPT1*) (Vesa et al., 1995). This enzyme, apparently located in lysosomes (Hellsten et al., 1996; Sleat et al., 1996), is thought to be involved in the catabolism of lipid-modified proteins, but its normal substrates have not been identified. The basis for the association of *PPT1* defects with the lysosomal accumulation of SAPs has not been determined.

The late-infantile form of NCL (CLN2) has a somewhat later onset, with clinical symptoms typically first appearing between 3 and 5 years of age. In this type of NCL, cognitive deficits are typically one of the earliest signs of the disease (Wisniewski et al., 1988). Cognitive decline is accompanied by seizures of various types and by severe motor deterioration. Affected children become blind due to retinal degeneration. As with the infantile disease, children with CLN2 eventually become vegetative. Death usually occurs between 7 and 10 years of age (Wisniewski et al., 1988). In this form of NCL, lysosomal storage bodies accumulate in cells throughout the central nervous system, as well as in many other cell types. The contents of the storage bodies have a distinctive ultrastructural appearance that is usually referred to as curvilinear profiles (Goebel and Sharp, 1998). A predominant molecular constituent of the storage bodies is the complete subunit c protein of mitochondrial inner membrane ATP synthase (Ezaki et al., 1995; Hall et al., 1991; Katz et al., 1997c; Kominami et al., 1992; Palmer et al., 1992). The gene that is defective in children with the late-infantile disease (the *CLN2* gene) apparently encodes a lysosomal peptidase (Sleat et al., 1997, 1999). How deficiencies in this enzyme lead to the specific accumulation of the subunit c protein or to the disease pathology are not understood.

The form of childhood NCL with the latest onset is the juvenile type (CLN3). In this disease, symptoms usually do not become apparent until 5–7 years of age. In the majority of cases, the initial symptom is visual loss due to retinal degeneration (Wisniewski et al., 1988). This is followed by the development of a variety of types of seizures and motor problems. Cognitive decline occurs more slowly in this form of NCL than in the earlier-onset types. Ultimately, affected children become vegetative, usually in their late teens. Death usually occurs in the late teens and early twenties. As in other forms of NCL, accumulation of lysosomal storage bodies occurs in neurons and many other cell types. The predominant ultrastructural appearance of the storage body contents is usually described as a fingerprint pattern. As in the late-infantile disorder, the storage bodies contain large amounts of the mitochondrial subunit c protein (Hall et al., 1991; Katz and Rodrigues, 1991; Palmer et al., 1992). The CLN3 gene that is defective in the juvenile disease encodes a protein that, based on the predicted amino acid sequence, is localized to a cellular membrane (International Batten Disease Consortium, 1995). Evidence obtained to date indicates that this protein is localized to either the mitochondria or lysosomes (Haskell et al., 1997; Jarvela et al., 1998; Kaczmarski et al., 1997; Katz et al., 1997a). As for the other forms of NCL, the mechanisms by which the gene defect leads to specific lysosomal storage or to the disease pathology are not known.

In addition to the more common forms of NCL, there are a number of rare variants that each appear to result from defects in different genes. At least eight distinct NCL diseases have been identified to date (Dyken and Wisniewski, 1995; Dyken, 1988; Klockars et al., 1996; Mole, 1999; Savukoski et al., 1998; Sharp et al., 1997). Among the more rare forms, it is not known whether the specific gene defects affect similar metabolic pathways to those affected by CLN1, CLN2, and CLN3 gene defects.

III. NATURALLY OCCURRING CEROID LIPOFUSCINOSIS IN ANIMALS AS MODELS FOR THE HUMAN DISORDERS

As indicated earlier, NCL-like disorders have been reported in a number of mammalian species. These include dogs, cats, cattle, sheep, and mice (Bildfell et al., 1995; Bronson et al., 1993; Chang et al., 1994; Goebel et al., 1988; Jarplid and Haltia, 1993; Jolly, 1995; Jolly et al., 1992a, 1992b; Jolly and Palmer, 1995; Koppang, 1988, 1992; Palmer et al., 1997; Tayler and Farrow, 1992; Weissenbock and Rossel, 1997). However, many of the disorders in animals were reported as isolated cases in veterinary patients, and only a small number of the animal diseases have been evaluated as potential models for the human disorders. The animal NCLs that have been most extensively studied will be described briefly.

A. English setter dogs

The first animal disease to be studied extensively as a potential model for the human NCLs occurs in English setters that originated in Norway. A disorder with similarities to human juvenile ceroid lipofuscinosis was first described in this breed in the late 1950s (Koppang, 1992). Affected dogs appear clinically normal early in life, but undergo a progressive neuronal degeneration that results in cognitive and motor decline. Neuronal degeneration is accompanied by blindness and seizures. The disease results in death at approximately 2 years of age. As with the human childhood NCLs, the disorder in the dogs was shown to be inherited in an autosomal recessive pattern (Koppang, 1988, 1992). Neuronal degeneration in the dogs is accompanied by a massive accumulation of autofluorescent lysosomal storage bodies in neurons and many other cell types (Katz et al., 1994; Koppang, 1988). The storage bodies were shown to be rich in the mitochondrial subunit c protein (Katz et al., 1994). On the basis of clinical, biochemical, and pathological features, the English setter disorder was long considered to be a good model for the juvenile disease in humans (Katz et al., 1994; Koppang, 1988, 1992).

With publication of the identity of the human *CLN3* gene in 1995 (International Batten Disease Consortium, 1995), it became possible to determine whether the canine disorder resulted from a defect in the canine ortholog of the gene that was involved in human juvenile NCL. Using the human gene sequence information, we determined the nucleotide sequences of the English setter ortholog of the *CLN3* gene and its cDNA in both affected and normal English setters. No sequence differences other than single-base substitutions in some of the introns were identified (Shibuya et al., 1998). Using an intragenic polymorphism as a marker, linkage analysis was performed with an informative English setter pedigree. Neither allele was specifically associated with the disease phenotype. Thus, defects in the canine ortholog of the *CLN3* gene were ruled out as the cause of the disease in this breed (Shibuya et al., 1998). This conclusion was confirmed by linkage analysis of a large English setter pedigree from Norway (Lingaas et al., 1998).

Elimination of a mutation at the *CLN3* locus as being the cause of NCL in the English setter led to the hypothesis that the canine disease might result from a defect in the ortholog of the *CLN2* gene. Using the human *CLN2* gene sequence information (Liu et al., 1998; Sleat et al., 1997), we determined the cDNA sequences of the canine *CLN2* ortholog in an affected and a normal English setter. No differences were detected (Figure 10.1). As with *CLN3*, a polymorphism was found in one of the canine *CLN2* introns that was informative for linkage analysis in an English setter family. The two alleles segregated independently from the disease in this family (Figure 10.2). Thus, *CLN2* was also eliminated as the locus for the English setter disease. Support for this conclusion is supported by the finding that the English setter disorder is not linked to canine orthologs of genes that map close to human *CLN2* (Lingaas et al., 1998). In addition, affected English setters showed normal levels of activity of the pepstatin-insensitive protease activity that has been attributed to the CLN2 protein (Sohar et al., 1999).

The clinical, biochemical, and pathological similarities between NCL in the English setter and the human late-infantile and juvenile disorders suggest that these dogs may be of some use as models for the human NCLs. However, elimination of defects in the *CLN2* and *CLN3* loci as causes for the canine disorder indicates that the English setter disease is not a close homolog of either of these human diseases. Thus, it is not apparent to what extent research with the English setter model would be applicable to human late-infantile or juvenile NCL. It is clear that there are a number of other types of NCL. The gene defect in the setters may be in an ortholog of one of the genes involved in these other, more rare forms.

B. South Hampshire sheep

An autosomal recessively inherited disease in South Hampshire sheep has been proposed as a model for some of the human NCLs. This model has been developed and studied extensively by R. D. Jolly and colleagues (Broom et al., 1998; Fearnley et al., 1990; Jolly, 1995; Jolly et al., 1980, 1988, 1992a, 1992b; Palmer et al., 1986a, 1986b, 1990). As in the human NCLs, the ovine disease is characterized by progressive atrophy of the retina and central nervous system. Generalized seizures do not typically occur in affected sheep, and when they do, it is usually when death is imminent. However, the sheep do develop localized tremors that have been interpreted as partial symmetrical seizures that do not generalize (Mayhew et al., 1985). As in the human disorders, affected sheep show massive accumulation of autofluorescent lysosomal storage bodies in neurons and many other cell types. On the basis of clinical symptoms, disease pathology, and storage body chemical composition, the sheep model most closely resembles the human juvenile disorder (Jolly et al., 1992b). However, genetic mapping studies indicate that the gene defect in the sheep is syntenic with a region of the human genome to which a late-infantile variant form of human NCL (CLN6) has been mapped (Broom et al., 1998).

The greatest contribution of the sheep model to our understanding of the NCLs was the discovery that the mitochondrial subunit c protein is a major constituent of the lysosomal storage bodies in these animals (Fearnley et al., 1990;

Figure 10.1. cDNA sequence of canine ortholog of the *CLN2* gene (top line), and predicted amino acid sequences of the canine (second line), human (third line) (Sleat et al., 1997), and mouse (fourth line) (Katz et al., 1999a) proteins. Hyphens in the human and mouse sequences indicate identity with the canine sequence. The N-terminal of the CLN2 protein that has been isolated from human tissues (residue 196) is indicated by an open arrow. The locations of the introns are indicated by filled arrows. Positions at which mutations resulting in single amino acid substitutions have been found in alleles of the gene associated with clinical ceroid lipofuscinosis are indicated by boxes (Sleat et al., 1999).

```
    cgcgtggtgtggcggtggaacataggttcatgtgatctgtcacatgacagtggatctccagaaggcaaa

    ATGAGACTCCGAACCTGCCTCCTAGGCTCCTTGCTCTGTGCCAGCAAATGCAGTTACAGCCCGAGCCAGAC
1   M   R   L   R   T   C   L   L   G   L   L   A   L   C   V   A   S   K   C   S   Y   S   P   E   P   D
    -   G   -   -   -   Q   A   -   -   -   F   -   -   -   -   -   L   S   G   -   -   T   -   N   -   -
    -   G   -   -   -   Q   A   R   -   -   -   -   -   -   -   V   I   -   G   -   -   -   -   -   -   -

    CAGCAGCGGACGCTGCCCCCAGGCTGGGTGTCCCTGGGCCGTGTAGATTCTGAGGAAGAGCTGAGTCTCACCTTTGCC
27  Q   Q   R   T   L   P   P   G   W   V   S   L   G   R   V   D   S   E   E   E   L   S   L   T   F   A
    -   R   -   -   -   -   -   -   -   -   -   -   -   -   -   -   A   -   P   -   -   -   -   -   -   -
    -   R   W   M   -   -   -   -   -   -   -   -   -   -   -   -   -   -   P   -   -   -   -   -   -   -

    CTGAGACAGCAGAACGTGAAAGATTGTCCAAGCTGTACAGGCTGTGTCGGATCCTGGCTCTCCTCATTATGAAAA
53  L   R   Q   Q   N   V   E   R   L   S   K   L   V   Q   A   V   S   D   P   P   G   S   P   H   Y   G   K
    -   -   -   -   -   -   -   -   -   -   -   -   -   -   -   -   -   -   -   -   -   S   -   Q   -   -   -
    -   K   -   -   -   -   -   R   -   L   -   -   -   E   -   -   -   -   -   -   -   S   -   Q   -   -   -

    TACCTGACCCTAGAGGATGTGGCTGAACTGGTCCGGCCATCACCACTGCCTCCGCACAGTCCAAAAATGGCTCTCA
79  Y   L   T   L   E   D   V   A   E   L   V   R   P   S   P   L   T   F   R   T   V   Q   K   W   L   S
    -   -   -   -   -   D   -   -   -   -   -   -   -   -   -   -   -   -   -   -   -   L   H   -   -   L
    -   -   -   -   -   -   -   -   -   -   -   -   -   Q   -   -   -   -   -   -   -   L   L   -   -   -

    GCAGCTGGAGCCCGGAACTGCCACTCGGTGACCACAAGACTTTCTGACTTGCTGGTCGAGTGTCGACAGGCGAGAA
105 A   A   G   A   R   N   C   H   S   V   T   T   Q   D   F   L   T   C   W   L   S   V   R   Q   A   E
    -   -   -   -   -   -   -   -   -   -   -   -   -   -   -   -   -   -   -   -   -   -   -   I   -   -
    -   -   -   -   -   Q   K   -   -   -   -   -   -   D   -   -   -   -   -   -   -   -   -   -   -   -

    CTGCTCCTCTCTGGGGCTGAGTTTCATCGCTATGTGGGGGACCTACAGAGATCCATGTTATAAGGTCCTACGTCCA
131 L   L   L   S   G   A   E   F   H   R   Y   V   G   G   P   T   E   I   H   V   I   R   S   L   R   P
    -   -   -   -   -   -   -   -   -   H   -   -   -   -   -   -   -   T   -   -   V   -   -   P   H   -
    -   -   -   -   -   -   -   -   -   -   -   -   -   -   -   -   -   K   T   -   -   -   -   P   H   -
                                                                                                    (continues)
```

```
    TACCAGCTCCCGAAGGCCTTGGCCCCTCATGTGGACTTTGTGGGCGGGCTGCACCGCTTCCCCCCCACATCATCCCTG
157 Y Q L P K A L A P H V D F V G G L H R F P P T S S L
    - - - - Q - - - - - - - - - - - - - H - - - - - - -
    - - - - Q - - - - - - - - - - - - - - - - -   - - P

    AGGCAACGCCCTGAGCCACAAGTGTCAGGGACTGTTGGCCTGCACCTGGGGGTGACCCCATCTGTGATCCGTCAGCGA
183 R Q R P E P Q V S G T V G L H L G V T P S V I R Q R
    - - - - - - - - T - - - - - - - - - - - - - - - K -
    - - - - - - - Q V - - - S - - - - - - - - L - - - -

    TACAACTTGACAGCACAAGATGTGGGCTCTGGCACAACCAACAACAGCCAAGCCTGTGCCCAGTTCCTGGAGCAGTAT
209 Y N L T A Q D V G S G T T N N S Q A C A Q F L E Q Y
    - - - - - - - - - - - - S - - - - - - - - - - - - -
    - - - - - K - - - - - - - - - - - - - - - - - - - -

    TTCCATGCATCAGACCTGGCTGAATTCATGCGCCTCTTTGGTGGGAACTTTGCACACCAGGCATCGGTAGCCCGTGTA
235 F H A S D L A E F M R L F G G N F A H Q A S V A R V
    - - D - - - - Q - - - - - - - - - - - - - - - - - -
    - - N - - - T - - - - - - - - S - T - - - - - - K -

    GTTGGACAGCAGGGCCGGGGCCGGGCTGGGATCGAGGCCAGTCTAGATGTGGAGTACCTGATGAGTGCCGGTGCCAAC
261 V G Q Q G R G R A G I E A S L D V E Y L M S A G A N
    - - - - - - - - - - - - - - - - - Q - - - - - - - -
    - - K - - - - - - - - - - - - - - - - - - - - - - -

    ATCTCCACCTGGGTCTACAGTAGCCCTGGCCGGCATGAGTCACAGGAGCCCTTCCTGCAGTGGCTCCTGCTGCTCAGT
287 I S T W V Y S S P G R H E S Q E P F L Q W L L L L S
    - - - - - - - - - - - - - G - - - - - - - M - - - -
    - - - - - - - - - - - - - A - - - - - - - - - - - -
```

Figure 10.1. *(continued)*

```
      AATGAGTCAGCCCTGCCACATGTGCACACTGTGAGCTATGGAGATGACGAGGACTCCCTCAGCAGCCTACATCCAG
313    N  E  S  A  L  P  H  V  H  T  V  S  Y  G  D  D  E  D  S  L  S  S  A  Y  I  Q
       -  -  -  -  -  -  -  -  -  -  -  -  -  -  -  -  -  -  -  -  -  -  -  I  -  -
       -  -  -  -  -  S  -  -  -  -  -  -  -  -  -  -  -  -  -  -  -  -  -  -  -  -

      CGGGTCAACACTGAGTTCATGAAGGCAGCGCTCGGGGTCTGACCCTGCTCTTTGCCTCAGGTGACAGTGGGGCTGGG
339    R  V  N  T  E  F  M  K  A  A  A  R  G  L  T  L  L  F  A  S  G  D  S  G  A  G
       -  -  -  -  -  L  -  -  -  -  -  -  -  -  -  -  -  -  -  -  -  -  -  T  -  -
       -  -  -  -  -  -  -  -  -  -  -  -  -  -  -  -  -  -  -  -  -  -  -  -  -  -

      TGTTGGTCTGTCTCTAGAAGACACCAGTTCCGTCCAGCTTCCGTCCTGCTCCAGCCCCTATGTCACCAGGTGGGAGGC
365    C  W  S  V  S  R  R  H  Q  F  R  P  S  F  P  A  S  S  P  Y  V  T  T  V  G
       -  -  -  -  -  -  -  G  -  -  -  -  -  T  -  -  -  -  -  -  -  -  -  -  -
       -  -  -  -  -  -  -  G  -  -  -  -  K  -  -  -  -  -  -  -  -  -  -  -  -

      ACATCCTTCCAGAATCCATTTCGAGTCACAACTGAGATTGTTGACTATATCAGTGGTGGCGGCTTCAGCAATGTGTTC
391    T  S  F  Q  N  P  F  R  V  T  T  E  I  V  D  Y  I  S  G  G  G  F  S  N  V  F
       -  -  -  -  -  E  -  -  L  I  -  -  N  -  -  -  -  -  -  -  -  -  -  -  -  -
       -  -  -  K  -  -  -  -  L  I  -  -  D  -  -  V  -  -  -  -  -  -  -  -  -  -

      CCACAGCCTTCATACCAGGAGGAAGCTGTGGTCCAGTTCCTGAGCTCCAGTCCCCACCTTAGTTATTTC
417    P  Q  P  S  Y  Q  E  E  A  V  V  Q  F  L  S  S  S  P  H  L  P  P  S  S  Y  F
       -  R  -  -  -  -  -  -  -  -  T  K  -  -  -  -  -  -  -  -  -  -  -  -  -  -
       -  R  -  -  -  -  -  -  -  -  A  -  -  -  -  -  K  -  -  -  -  -  -  -  -  -

      AATGCCAGTGGCCGTGCCTATCCAGACGTGGCTCTCCGATGGCTACTGGGTGGTCAGTAACAGCGTGCCCATT
443    N  A  S  G  R  A  Y  P  D  V  A  L  S  D  G  Y  W  V  V  S  N  S  V  P  I
       -  -  -  -  -  -  -  -  -  -  -  -  -  -  -  -  -  -  -  -  -  R  -  -  -
       -  -  -  -  -  -  -  -  -  -  -  -  -  -  -  -  -  -  -  -  -  M  -  -  -
```

(*continues*)

```
                   ↓
469  CCATGGGTGTCTGGCACCTCGGCTTCTACTCCAGTGTTTGGTGGATCCTATCCTGATAAATGAACATAGACTCCTC
     P  W  V  S  G  T  S  A  S  T  P  V  F  G  G  I  L  S  L  I  N  E  H  R  L
     -  -  -  -  -  -  -  -  -  -  -  -  -  -  -  -  -  -  -  -  -  -  -  I  -
     -  -  -  -  -  -  -  -  -  -  -  -  -  -  -  -  -  -  -  -  -  -  -  I  -

                                                                    ↓
495  AGTGGCCTCCCTCCTCTTGGCTTTCTCAACCCAAGCTCTACCAGCAGCGTGGGCAGGAGACTCTTTGATGTAACCCGT
     S  G  L  P  P  L  G  F  L  N  P  R  L  Y  Q  Q  R  G  A  G  L  F  D  V  T  R
     -  -  R  -  -  -  -  -  -  -  -  -  -  -  -  -  -  -  H  -  -  -  -  -  -  -
     N  -  R  -  -  -  -  -  -  -  -  -  -  -  -  -  -  -  H  -  T  -  -  -  -  H

521  GGCTGCCATGAATCTGTCTGAATGAAGAGGTGCAGGGTCAGGGTTCTGCTCTGGCCCTGGCTGGATCCTGTGACA
     G  C  H  E  S  C  L  N  E  E  V  Q  G  Q  G  F  C  S  G  P  G  W  D  P  V  T
     -  -  -  -  -  -  -  -  D  -  -  -  -  -  -  -  -  -  -  -  -  -  -  -  -  -
     -  -  -  -  -  -  -  -  -  -  -  -  E  -  -  -  -  -  -  -  -  -  -  -  -  -

547  GGCTGGGAACACCCAACTTCCCAGCTCTGCTGAAGGCACTAATCAAACCTTGAcccttcctgctggacactgaac
     G  W  G  T  P  N  F  F  P  A  L  L  K  A  L  I  K  P  *
     -  -  -  -  -  -  -  -  -  -  -  -  -  -  -  T  -  L  N  -  *
     -  -  -  -  -  -  -  -  -  -  -  -  -  -  -  T  -  L  N  -  *

     tggtccctgcccatagctggtggctggtcttgcactgtcgactgtcggaaggcctgttgaaccctcaaccataga
     ctgcaacagacagcttattcccctaaccctgaaatgctgtgagctgacttgactcctaaccctgactcctccatcat
     gctcaggtcttcctactccg
```

Figure 10.1. (*continued*)

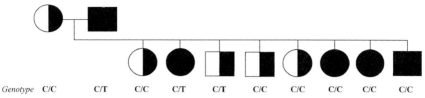

Figure 10.2. Pedigree of an English setter family showing the pattern of inheritance of two alleles of the *CLN2* gene. The two alleles are distinguished by a single nucleotide difference (T or C). Affected dogs are indicated by filled symbols and phenotypically normal obligate heterozygotes are indicated by half-filled symbols. There was no association between CLN2 genotype and disease status.

Palmer *et al.*, 1986a, 1989, 1990). This observation led to the demonstration that the same small hydrophobic protein is abundant in the storage bodies that accumulate in most forms of human NCL other than the infantile type (Ezaki *et al.*, 1995, 1996; Hall *et al.*, 1991; Katz *et al.*, 1995, 1997c; Kominami *et al.*, 1992; Palmer *et al.*, 1992; Rowan and Lake, 1995; Tyynela *et al.*, 1997b; Wisniewski *et al.*, 1995). Determining why this specific protein accumulates in the NCLs may be the key to understanding how mutations in a variety of different genes result in similar disorders. In this respect, the sheep should be a good model for elucidating the mechanisms involved in the specific accumulation of the subunit c protein.

C. Tibetan terrier dogs

Ceroid lipofuscinosis has been described in a number of dog breeds in addition to the English setters (Appleby *et al.*, 1982; Cummings and deLahunta, 1977; Goebel *et al.*, 1988; Jolly, 1995; Palmer *et al.*, 1997; Rac and Giesecke, 1975; Tayler and Farrow, 1992). However, the Tibetan terrier is the only other breed that has been studied in sufficient detail for these animals to be evaluated as a model for the human disorders. Like the human childhood-onset NCLs, ceroid lipofuscinosis in Tibetan terriers is inherited in an autosomal recessive pattern (Riis *et al.*, 1992). The first clinical symptom to become apparent is night blindness, which can be detected with electrorintography as early as 7 weeks of age (Riis *et al.*, 1992). As in the human disorders, vision loss is due primarily to a progressive retinal degeneration. Behavioral changes and loss of training begin to appear between 4 and 6 years of age. Mild ataxia develops at 8–9 years. Unlike the human childhood disorders, neither seizures nor shortening of life span have been reported in these animals. Clinically, the Tibetan terrier disease has a later onset and a milder course that the human NCLs. The disease in Tibetan terriers, like that in humans and other animals, is accompanied by accumulation of autofluorescent lysosomal storage bodies in neurons and other cell types (Alroy *et al.*, 1992; Riis *et al.*, 1992). Storage body accumulation is accompanied by neuronal degeneration in both the retina and the brain. However, brain atrophy appears to be much more mild than in the human NCLs. As in the human late-infantile and juvenile NCLs, the lysosomal storage bodies in the Tibetan terriers are rich in the mitochondrial subunit c protein (Jolly et al., 1992b).

To determine whether the Tibetan terrier is homologous to one of the human NCLs, it will be necessary to identify or at least map the gene defect in these dogs. Studies have been initiated to determine whether affected Tibetan terriers have a mutation in the canine ortholog of either the *CLN2* or *CLN3* gene. In addition, DNA samples are being collected from appropriate Terrier families to map the disorder on the evolving canine genome map (Mellersh et al., 2000). Until these genetic analyses are completed, it will not be known whether the Tibetan terrier disease is a suitable model for one of the human NCLs.

D. *mnd* Mice

A naturally occurring mutation resulting in motor neuron degeneration in a substrain of C57BL/6 mice was first described almost 15 years ago (Messer and Flaherty, 1986). Initially, the disease in this mouse strain (since designated *mnd*) was proposed as a model for amyotropic lateral sclerosis (ALS) (Messer et al., 1987). However, histological examination of neural tissues led to the discovery that the disease is characterized by a progressive accumulation of autofluorescent intracellular inclusions, similar to those that accumulate in the NCLs (Messer and Plummer, 1993). Subsequently, affected mice were found to undergo retinal degeneration (Chang et al., 1994; Messer et al., 1993), and the intracellular inclusions that accumulate in cells were shown to contain the mitochondrial subunit c protein (Bronson et al., 1993). Thus, the *mnd* strain was proposed as a model of NCL (Bronson et al., 1993; Chang et al., 1994). Recently, the mutation causing the disease in *mnd* mice was found to be a 1-bp insertion in the ortholog of a gene that when defective in humans causes progressive epilepsy with mental retardation (EPMR) (Ranta et al., 1999). The human disease, now classified as a new NCL subtype (CLN8), is an autosomal recessive disorder characterized by an onset of generalized seizures between 5 and 10 years of age and subsequent progressive metal retardation. The function of the protein encoded by the *CLN8* gene is not known, but based on the predicted amino acid sequence, it is likely to be a transmembrane protein (Ranta et al., 1999). Abnormalities have been reported in mitochondria-associated membranes of *mnd/mnd* mice (Vance et al., 1997), which suggests that the *CLN8* gene might encode a mitochondrial membrane protein. Dietary supplementation with carnitine slowed the disease progression in *mnd/mnd* mice and prolonged their life spans (Katz et al., 1997b), suggesting that a similar treatment may be beneficial for humans with the CLN8 disorder.

E. *nclf* Mice

A mouse strain with a phenotype very similar to that of the *mnd* strain has been described and proposed as another potential animal model for NCL (Bronson et al., 1998). This mouse strain has been designated *nclf*. The disorder in these mice is inherited as an autosomal recessive trait. Homozygous *nclf* mice develop progressive retinal atrophy early in life and become paralyzed at about 9 months of age. Lysosomal storage bodies with staining and ultrastructural properties similar to those of the storage material in the human NCLs were found to accumulate in neurons and many other cell types. The genetic defect in the *nclf* mice has been mapped to a region of mouse chromosome 9 that is syntenic to human chromosome 15q21, where the gene that is defective in the CLN6 form of human NCL has been mapped (Sharp et al., 1997). Thus, the *nclf* strain is likely to be a suitable model for the latter disorder.

IV. ANIMAL MODELS CREATED THROUGH MOLECULAR GENETIC MANIPULATION

With the development of technology for molecular manipulation of specific genes in mice, it has become possible to generate mouse strains in which the murine orthologs of genes involved in human inherited diseases are disrupted (Capechi, 1994). This technique, designated targeted gene replacement, has been employed to generate mouse models for a number of inherited human disorders (Bahn et al., 1999; Bourdeau et al., 1999; Leheste et al., 1999; Rudmann and Durham, 1999; Taketo, 1999; Willecke et al., 1999; Wong et al., 1999). Once the genes involved in the human NCLs were identified, it became possible to employ this technology to develop mouse models for those forms of NCL for which no naturally occurring mutant mouse strains have been identified. Targeted gene replacement has been or is currently being used to develop mouse models of the human CLN2 and CLN3 disorders.

A. Mouse gene knockout models of juvenile NCL (*CLN3*)

The mouse ortholog of the *CLN3* gene consists of 15 exons and with the intervening introns is over 14 kb in length (Katz et al., 1999b). The mouse gene is designated *Cln3*. Gene targeting has been used to generate two mouse strains in which the *Cln3* gene has been disrupted by replacement of portions of the normal gene with a neomycin-resistance gene (Figure 10.3) (Greene et al., 1999; Katz et al., 1999b). In both strains, there is widespread intracellular accumulation of autofluorescent lysosomal storage bodies with ultrastructural characteristics similar to those that accumulate in human juvenile NCL (Katz et al.,

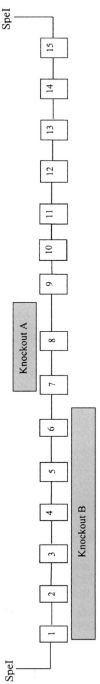

Figure 10.3. Regions of the *Cln3* gene that have been knocked out in two mouse strains are shown in filled boxes above and below the schematic diagram of the *Cln3* gene (not drawn to scale). The mouse gene contains 15 exons. Most of exon 7 and all of exon 8 are deleted in knockout A (Katz *et al.*, 1999b), whereas most of exon 1 and exons 2-6 are deleted in knockout B (Greene *et al.*, 1999).

1999b; Mitchison et al., 1999). In one strain, behavioral abnormalities including hyperactivity and hind limb stiffness have been observed (Katz et al., 1999b). Premature death between 4 and 10 months of age occurred in many of the mice from the latter strain (Katz et al., 1999b). There was substantial phenotypic variability these animals. In the mice characterized by Katz and colleagues to date, the *Cln3* mutation was on a mixed C57BL/6 and 129SV/J background (Katz et al., 1999b). This mixed background was probably responsible for the phenotypic variability. Backcrosses are currently being performed to place the *Cln3* mutation on a pure C57BL/6 background. When this has been accomplished, further phenotypic characterization will be undertaken.

In the *Cln3* knockout strain generated by Greene and colleagues (Greene et al., 1999), a more detailed phenotypic characterization was carried out in mice with the mutation on a mixed 129SV–NIH Black Swiss genetic background (Mitchison et al., 1999). The storage bodies in these animals stained with an antibody directed against the mitochondrial subunit c protein, a major constituent of the storage material in human juvenile NCL. In these animals, there were reduced numbers of specific neuronal cell types in the cerebral cortex compared to normal control mice. In other areas of the brain, there was neuronal hypertrophy. Elevated activity levels of the lysosomal peptidase that is absent in human CLN2 patients were also detected in the affected mice. The latter observation most likely reflects an overall increase in lysosomal enzymes in these animals.

On the basis of the characterizations carried out to date, it appears that both *Cln3* mutant mouse strains will be useful in studying the mechanisms underlying disease pathology in juvenile NCL and for screening potential therapeutic interventions.

B. Mouse gene knockout model of late-infantile NCL

Identification of the gene that is defective in human children with late-infantile NCL (Sleat et al., 1997) enabled work to begin on constructing a mouse gene knockout model for this disorder. The first step in this process was to characterize the mouse ortholog of the *CLN2* gene (Katz et al., 1999a). The mouse *Cln2* gene had the same genomic organization as its human ortholog, although the intron sizes varied somewhat between the species. The mouse gene spans more than 6 kb and consists of 13 exons separated by introns ranging in size from 111 to 1259 bp (Figure 10.4). The mouse cDNA contains an open reading frame that predicts a protein product of 562 amino acids. The mouse and human coding regions are 86% and 88% identical at the nucleic acid and amino acid levels, respectively, with one less codon in the mouse than in the human cDNA (Figure 10.1). The *Cln2* gene maps to a region of mouse chromosome 7 that corresponds to human chromosome 11p15 where the *CLN2* gene has been mapped.

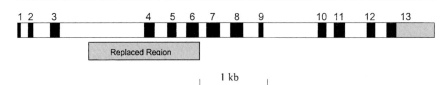

Figure 10.4. Region of the *Cln2* gene that has been replaced by the neomycin resistance gene in the *Cln2* knockout mouse strain that is under development. The *Cln2* gene (drawn to scale) contains 13 exons (shown by filled boxes). Exons 4–6 are deleted in the knockout construct.

Once the *Cln2* gene had been characterized, a targeting construct was made to disrupt this gene in mouse embryonic stem (ES) cells. After incorporation of the targeting construct, an ES clone was identified in which three of the *Cln2* exons were replaced by a neomycin resistance gene (Figure 10.4). The ES cells containing this knockout allele were incorporated into blastocyst-stage embryos and transferred into pseudo-pregnant recipient foster mothers. The embryos developed into mice that were chimeras containing a significant fraction of their cells derived from the modified ES cell line. To complete the development of the *Cln2* knockout model, the chimeric mice are being bred to normal animals, and the offspring will be screened for heterozygosity at the *Cln2* locus. Heterozygous mice will then be bred to obtain animals that are homozygous for the knockout allele. The homozygous knockout animals will be monitored for the appearance of phenotypic features of NCL.

V. FUTURE DIRECTIONS

None of the animal models identified or developed to date replicate the human NCL disorders perfectly. Nonetheless, they have been and will continue to be valuable tools in studies to determine the mechanisms that underlie the human disease pathologies. The availability of these animal models will also hasten the development of effective treatment interventions for the human NCLs. Screening of potential therapies can be done relatively quickly and at modest cost using animal models, particularly the mouse models. The most promising therapies can then be evaluated in human clinical trials. For those afflicted with the NCLs and their families, the time for hope is short, as the neurological degeneration progresses relentlessly. The goal in developing suitable animal models is to provide tools that will be keys to discovery of effective therapeutic interventions for the NCLs that will halt or at least delay the neurological degeneration associated with these disorders.

Acknowledgments

Original work conducted in the authors' laboratories was supported by the Batten Disease Support and Research Association, the Children's Brain Diseases Foundation, and by National Institutes of Health grants NS30155 and NS38987.

References

Alroy, J., Schelling, S. H., Thalhammer, J. G., Raghavan, S. S., Natowicz, M. R., Prence, E. M., and Orgad, U. (1992). Adult onset lysosomal storage disease in a Tibetan terrier: Clinical, morphological and biochemical studies. *Acta Neuropathol.* **84,** 658–663.

Appleby, E. C., Longstaffe, J. A., and Bell, F. R. (1982). Ceroid-lipofuscinosis in two saluki dogs. *J. Comp. Pathol.* **92,** 375–380.

Bahn, A. K., Mizoguchi, E., Smith, R. N., and Mizoguchi, A. (1999). Colitis in transgenic and knockout animals as models of human inflammatory bowel disease. *Immunol. Rev.* **169,** 195–207.

Bildfell, R., Matwichuk, C., Mitchell, S., and Ward, P. (1995). Neuronal ceroid-lipofuscinosis in a cat. *Vet. Pathol.* **32,** 485–488.

Bourdeau, A., Dumont, D. J., and Letarte, M. (1999). A murine model of hereditary hemorrhagic telangiectasia. *J. Clin. Invest.* **104,** 1343–1351.

Boustany, R. N., Alroy, J., and Kolodny, E. H. (1988). Clinical classification of neuronal ceroid lipofuscinosis subtypes. *Am. J. Med. Genet.* **Suppl. 5,** 47–58.

Bronson, R. T., Lake, B. D., Cook, S., Taylor, S., and Davisson, M. T. (1993). Motor neuron degeneration of mice is a model of neuronal ceroid lipofuscinosis (Batten's disease). *Ann. Neurol.* **33,** 381–385.

Bronson, R. T., Donahue, L. R., Johnson, K. R., Tanner, A., Lane, P. W., and Faust, J. R. (1998). Neuronal ceroid lipofuscinosis (*nclf*), a new disorder of the mouse linked to chromosome 9. *Am. J. Med. Genet.* **77,** 289–297.

Broom, M. F., Zhou, C. M., Broom, J. E., Barwell, K. J., Jolly, R. D., and Hill, D. F. (1998). Ovine ceroid lipofuscinosis: A large animal model syntenic with the human neuronal ceroid lipofuscinosis variant CLN6. *J. Med. Genet.* **35,** 717–721.

Capechi, M. R. (1994). Targeted gene replacement. *Sci. Am.* **270,** 52–59.

Chang, B., Bronson, R. T., Hawes, N. L., Roderick, T. H., Peng, C., Hageman, G. S., and Heckenlively, J. R. (1994). Retinal degeneration in motor neuron degeneration: A mouse model for ceroid-lipofuscinosis. *Invest. Ophthalmol. Vis. Sci.* **35,** 1071–1076.

Cummings, J. F., and deLahunta, A. (1977). An adult case of canine ceroid-lipofuscinosis. *Acta Neuropathol.* **39,** 43–51.

Dyken, P. R. (1988). Reconsideration of the classification of the neuronal ceroid-lipofuscinoses. *Am. J. Med. Genet.* **Suppl. 5,** 69–84.

Dyken, P., and Wisniewski, K. (1995). Classification of the neuronal ceroid-lipofuscinoses: Expansion of the atypical forms. *Am. J. Med. Genet.* **57,** 150–154.

Ezaki, J., Wolfe, L. S., Higuti, T., Ishidoh, K., and Kominami, E. (1995). Specific delay of degradation of mitochondrial ATP synthase subunit c in late infantile neuronal ceroid lipofuscinosis (Batten disease). *J. Neurochem.* **64,** 733–741.

Ezaki, J., Wolfe, L. S., and Kominami, E. (1996). Specific delay in degradation of mitochondrial ATP synthase subunit c in late infantile neuronal ceroid lipofuscinosis is derived from cellular proteolytic dysfunction rather than structural alteration of subunit c. *J. Neurochem.* **67,** 1677–1687.

Fearnley, I. M., Walker, J. E., Martinus, R. D., Jolly, R. D., Kirkland, K. B., Shaw, G. J., and Palmer, D. N. (1990). The sequence of the major protein stored in ovine ceroid-lipofuscinosis is identical with that of the dicyclohexylcarbodiimide-reactive proteolipid of mitochondrial ATP synthase. *Biochem. J.* **268,** 751–758.

Goebel, H. H., and Sharp, J. D. (1998). The neuronal ceroid-lipofuscinoses: Recent advances. *Brain Pathol.* **8,** 151–162.

Goebel, H. H., Bilzer, T., Dahme, E., and Malkush, F. (1988). Morphological studies in canine (dalmation) neuronal ceroid-lipofuscinosis. *Am. J. Med. Genet.* **Suppl. 5,** 127–140.

Greene, N. D. E., Bernard, D. L., Taschner, P. E. M., Lake, B. D., de Vos, N., Breuning, M. H., Gardiner, R. M., Mole, S. E., Nussbaum, R. L., and Mitchison, H. M. (1999). A murine model for juvenile NCL: Gene targeting of mouse *Cln3*. *Mol. Genet. Matabol.* **66,** 309–313.

Hall, N. A., Lake, B. D., Dewji, N. N., and Patrick, A. D. (1991). Lysosomal storage of subunit c of mitochondrial ATP synthase in Batten's disease (ceroid-lipofuscinosis). *Biochem. J.* **275,** 269–272.

Haskell, R. E., Derken, T. A., and Davidson, B. L. (1997). Intracellular localization of the Batten gene product, CLN3. *Soc. Neurosci.* **23,** 860.

Hellsten, E., Vesa, J., Olkkonen, V. M., Jalanko, A., and Peltonen, L. (1996). Human palmitoyl protein thioesterase: Evidence for lysosomal targeting and disturbed cellular routing in infantile neuronal ceroid lipofuscinosis. *EMBO J.* **15,** 5240–5245.

International Batten Disease Consortium (1995). Isolation of a novel gene underlying Batten disease, CLN3. *Cell.* **82,** 949–957.

Jarplid, B., and Haltia, M. (1993). An animal model for the infantile type of neuronal ceroid-lipofuscinosis. *J. Inher. Metab. Dis.* **16,** 274–277.

Jarvela, I., Sainio, M., Rantamki, T., Olkkonen, V. M., Carpen, O., Peltonen, L., and Jalanko, A. (1998). Biosynthesis and intracellular targeting of the CLN3 protein defective in Batten disease. *Hum. Molec. Genet.* **7,** 85–90.

Jolly, R. D. (1995). Comparative biology of the neuronal ceroid-lipofuscinoses (NCL): An overview. *Am. J. Med. Genet.* **57,** 307–311.

Jolly, R. D., and Palmer, D. N. (1995). The neuronal ceroid-lipofuscinoses (Batten disease): Comparative aspects. *Neuropathol. Appl. Neurobiol.* **21,** 50–60.

Jolly, R. D., Janmaat, A., West, D. M., and Morrison, I. (1980). Ovine ceroid-lipofuscinosis I: A model for Batten's disease. *Neuropathol. Appl. Neurobiol.* **6,** 195–206.

Jolly, R. D., Shimada, A., Craig, A. S., Kirkland, K. B., and Palmer, D. N. (1988). Ovine ceroid-lipofuscinosis II: Pathologic changes interpreted in light of biochemical observations. *Am. J. Med. Genet.* **Suppl. 5,** 159–170.

Jolly, R. D., Gibson, A. J., Healy, P. J., Slack, P. M., and Birtles, M. J. (1992a). Bovine ceroid-lipofuscinosis: Pathology of blindness. *N. Z. Vet. J.*

Jolly, R. D., Martinus, R. D., and Palmer, D. N. (1992b). Sheep and other animals with ceroid-lipofuscinosis: Their relevance to Batten disease. *Am. J. Med. Genet.* **42,** 609–614.

Kaczmarski, W., Kida, E., Lach, A., Rubenstein, R., Zhong, N., and Wisniewski, K. (1997). Expression studies of CLN3 protein. *Neuropediatrics* **28,** 33–36.

Katz, M. L., and Rodrigues, M. (1991). Juvenile ceroid-lipofuscinosis: Evidence for methylated lysine in storage body protein. *Am. J. Pathol.* **138,** 323–332.

Katz, M. L., Christianson, J. S., Norbury, N. E., Gao, C., Siakotos, A. N., and Koppang, N. (1994). Lysine methylation of mitochondrial ATP synthase subunit c stored in tissues of dogs with hereditary ceroid-lipofuscinosis. *J. Biol. Chem.* **269,** 9906–9911.

Katz, M. L., Gao, C., Tompkins, J. A., Chin, D. T., and Bronson, R. T. (1995). Mitochondrial ATP synthase subunit c stored in hereditary ceroid-lipofuscinosis contains trimethyllysine. *Biochem. J.* **310,** 887–892.

Katz, M. L., Gao, C., Prabhakaram, M., Shibuya, H., Liu, P., and Johnson, G. S. (1997a). Immunochemical localization of the Batten disease (CLN3) protein in retina. *Invest. Ophthalmol. Vis. Sci.* **38,** 2373–2384.

Katz, M. L., Rice, L. M., and Gao, C. (1997b). Dietary carnitine supplements slow disease progression in a putative mouse model for hereditary ceroid-lipofuscinosis. *J. Neurosci. Res.* **50,** 123–132.

Katz, M. L., Siakotos, A. N., Gao, Q., Freiha, B., and Chin, D. T. (1997c). Late-infantile ceroid-lipofuscinosis: Lysine methylation of mitochondrial subunit c from lysosomal storage bodies. *Biochim. Biophys. Acta* **1361,** 66–74.

Katz, M. L., Liu, P., Grob-Nunn, S., Shibuya, H., and Johnson, G. S. (1999a). Characterization and chromosomal mapping of a mouse homolog of the late-infantile ceroid-lipofuscoinosis gene CLN2. *Mamm. Genome.* **10,** 1050–1053.

Katz, M. L., Shibuya, H., Liu, P., Kaur, S., Gao, C., and Johnson, G. S. (1999b). A mouse gene knockout model for juvenile ceroid-lipofuscinosis. *J. Neurosci. Res.* **57,** 551–556.

Klockars, T., Savukoski, M., Isosomppi, J., Laan, M., Jarvela, I., Petrukhin, K., Palotie, A., and Peltonen, L. (1996). Efficient construction of a physical map by fiber-fish of the CLN5 region: Refined assignment and long-range contig covering the critical region on 13q22. *Genomics* **35,** 71–78.

Kominami, E., Ezaki, J., Muno, D., Ishido, K., Ueno, T., and Wolfe, L. S. (1992). Specific storage of subunit c of mitochondrial ATP synthase in lysosomes of neuronal ceroid lipofuscinosis (Batten's disease). *J. Biochem.* **111,** 278–282.

Koppang, N. (1988). The English setter with ceroid-lipofuscinosis: A suitable model for the juvenile type of ceroid-lipofuscinosis. *Am. J. Med. Genet.* **Suppl. 5,** 117–126.

Koppang, N. (1992). English setter model and juvenile ceroid-lipofuscinosis in man. *J. Med. Genet.* **42,** 594–599.

Leheste, J. R., Rolinski, B., Vorum, H., Nykjaer, A., Jacobsen, C., Aucouturier, P., Moskaug, J. O., Otto, A., Christensen, E. I., and Willnow, T. E. (1999). Megalin knockout mice as an animal model of low molecular weight proteinuria. *Am. J. Pathol.* **155,** 1361–1370.

Lingaas, F., Aarskaug, T. M. S., Bjerkas, I., Grimholt, U., Moe, L., Juneja, R. K., Wilton, A. N., Galibert, F., Holmes, N. G., and Dolf, G. (1998). Genetic markers linked to neuronal ceroid lipofuscinosis in English setter dogs. *Animal Genet.* **29,** 371–376.

Liu, C., Sleat, D. E., Donnelly, R. J., and Lobel, P. (1998). Structural organization and sequence of CLN2, the defective gene in classical late infantile neuronal ceroid lipofuscinosis. *Genomics* **50,** 206–212.

Mayhew, I. G., Jolly, R. D., Pickett, B. T., and Slack, P. M. (1985). Ceroid-lipofuscinosis (Batten's disease): Pathogenesis of blindness in the ovine model. *Neuropathol. Appl. Neurobiol.* **11,** 273–290.

Mellersh, C. S., Hitte, C., Richman, M., Vignaux, F., Priat, C., Jouquand, S., Werner, P., Adre, C., DeRose, S., Patterson, D. F., Ostrander, E. A., and Galibert, F. (2000). An integrated linkage-radiation hybrid map of the canine genome. *Mamm. Genome* **11,** 120–130.

Messer, A., and Flaherty, L. (1986). Autosomal dominance in a late-onset motor neuron disease in the mouse. *Ann. Neurol.* **33,** 381–385.

Messer, A., and Plummer, J. (1993). Accumulating autofluorescent material as a marker for early changes in the spinal cord of the Mnd mouse. *Neuromusc. Disord.* **3,** 129–134.

Messer, A., Strominger, N. L., and Mazurkiewicz, J. E. (1987). Histopahtology of the late-onset motor neuron degeneration (Mnd) mutant in the mouse. *J. Neurogenet.* **4,** 201–213.

Messer, A., Plummer, J., Wong, V., and LaVail, M. M. (1993). Retinal degeneration in motor neuron degeneration (mnd) mutant mice. *Exp. Eye Res.* **57,** 637–641.

Mitchison, H. M., Bernard, D. J., Greene, N. D., Cooper, J. D., Juniad, M. A., Pullarkat, R. K., deVos, N., Bruening, M. H., Owens, J. W., Mobley, W. C., Gardiner, R. M., Lake, B. D., Taschner, P. E., and Nussbaum, R. L. (1999). Targeted disruption of the Cln3 gene provides a mouse model for Batten disease. *Neurobiol. Dis.* **6,** 321–334.

Mole, S. E. (1999). Batten's disease: Eight genes and still counting?. *Lancet* **354,** 443–445.

Palmer, D. N., Barns, G., Husbands, D. R., and Jolly, R. D. (1986a). Ceroid lipofuscinosis in sheep. II. The major component of the lipopigment in liver, kidney, pancreas, and brain is low molecular weight protein. *J. Biol. Chem.* **261,** 1773–1777.

Palmer, D. N., Husbands, D. R., Winter, P. J., Blunt, J. W., and Jolly, R. D. (1986b). Ceroid lipofuscinosis in sheep. I. Bis(monoacylglycero)phosphate, dolichol, ubiquinone, phospholipids, fatty acids, and fluorescence in liver lipopigment lipids. *J. Biol. Chem.* **261,** 1766–1772.

Palmer, D. N., Martinus, R. D., Barns, G., Reeves, R. D., and Jolly, R. D. (1988). Ovine ceroid-lipofuscinosis: Lipopigment composition is indicative of a lysosomal proteinosis. *Am. J. Med. Genet.* **Suppl. 5,** 141–158.

Palmer, D. N., Martinus, R. D., Cooper, S. M., Midwinter, G. G., Reid, J. C., and Jolly, R. D. (1989). Ovine ceroid-lipofuscinosis. The major lipopigment protein and the lipid-binding subunits of mitochondrial ATP synthase have the same NH_2-terminal sequence. *J. Biol. Chem.* **264,** 5736–5740.

Palmer, D. N., Fearnley, I. M., Medd, S. M., Walker, J. E., Martinus, R. D., Bayliss, S., Hall, N. A., Lake, B. D., Wolfe, L. S., and Jolly, R. D. (1990). Lysosomal storage of the DCCD-reactive proteolipid subunit of mitochondrial ATP synthase in human and ovine ceroid-lipofuscinosis. *In* "Lipofuscin and Ceroid Pigments" (E.A. Porta, ed.), pp. 211–223. Plenum Press, New York.

Palmer, D. N., Fearnley, I. M., Walker, J. E., Hall, N. A., Lake, B. D., Wolfe, L. S., Haltia, M., Martinus, R. D., and Jolly, R. D. (1992). Mitochondrial ATP synthase subunit c storage in the ceroid-lipofuscinoses (Batten disease). *Am. J. Med. Genet.* **42,** 561–567.

Palmer, D. N., Tyynela, J., van Mil, H. C., Westlake, J. V., and Jolly, R. D. (1997). Accumulation of sphingolipid activator proteins (SAPs) A and D in granular osmiophilic deposits in miniature Schnauzer dogs with ceroid lipofuscinosis. *J. Inher. Metab. Dis.* **20,** 74–84.

Rac, R., and Giesecke, P. R. (1975). Lysosomal storage disease in Chihuahuas. *Austral. Vet. J.* **51,** 403–404.

Ranta, S., Zhang, Y., Ross, B., Lonka, L., Takkunen, E., Messer, A., Sharp, J., Wheeler, R., Kusumi, K., Mole, S., Liu, W., Soares, M. B., Bonaldo, M. F., Hirvasniemi, A., de la Chapelle, A., and Lehesjoki, A. E. (1999). The neuronal ceroid lipofuscinoses in human EPMR and mnd mutant mice are associated with mutations in CLN8. *Nat. Genet.* **23,** 233–236.

Riis, R. C., Cummings, J. F., Loew, E. R., and de Lahunta, A. (1992). Tibetan terrier model of canine ceroid lipofuscinosis. *Am. J. Med. Genet.* **42,** 615–621.

Rowan, S. A., and Lake, B. D. (1995). Tissue and cellular distribution of subunit c of ATP synthase in Batten disease (neuronal ceroid-lipofuscinosis). *Am. J. Med. Genet.* **57,** 172–176.

Rudmann, D. G., and Durham, S. K. (1999). Utilization of genetically altered animals in the pharmaceutical industry. *Toxicol. Pathol.* **27,** 111–114.

Savukoski, M., Klockars, T., Holmberg, V., Santavuori, P., Lander, E. S., and Peltonen, L. (1998). CLN5, a novel gene encoding a putative transmembrane protein mutated in Finnish variant late infantile neuronal ceroid lipofuscinosis. *Nat. Genet.* **19,** 286–288.

Sharp, J. D., Wheeler, R. B., Lake, B. D., Savukoski, M., Jarvela, I. E., Peltonen, L., Gardiner, R. M., and Williams, R. E. (1997). Loci for classical and a variant late infantile neuronal ceroid lipofuscinosis map to chromosomes 11p15 and 15q21-23. *Hum. Mol. Genet.* **6,** 591–595.

Shibuya, H., Liu, P., Katz, M. L., Siakotos, A. N., Nonneman, D. J., and Johnson, G. J. (1998). Coding sequence and exon/intron organization of the canine CLN3 (Batten disease) gene and its exclusion as the locus for ceroid lipofuscinosis in English setter dogs. *J. Neurosci. Res.* **52,** 268–275.

Sleat, D. E., Sohar, I., Lackland, H., Majercak, J., and Lobel, P. (1996). Rat brain contains levels of mannose-6-phosphorylated glycoproteins including lysosomal enzymes and palmitoyl-protein thioesterase, an enzyme implicated in infantile neuronal lipofuscinosis. *J. Biol. Chem.* **271,** 19191–19198.

Sleat, D. E., Donnelly, R. J., Lackland, H., Liu, C., Sohar, I., Pullarkat, R. K., and Lobel, P. (1997).

Association of mutations in a lysosomal protein with classical late-infantile neuronal ceroid lipofuscinosis. *Science* **277**, 1802–1805.

Sleat, D. E., Gin, R. M., Sohar, I., Wisniewski, K., Sklower-Brooks, S., Pullarkat, R. K., Palmer, D. N., Lerner, T. J., Boustany, R., Uldall, P., Siakotos, A. N., Donnelly, R. J., and Lobel, P. (1999). Mutational analysis of the defective protease in classical late-infantile neuronal ceroid lipofuscinosis, a neurodegenerative lysosomal storage disorder. *Am. J. Hum. Genet.* **64**, 1511–1523.

Sohar, I., Sleat, D. E., Jadot, M., and Lobel, P. (1999). Biochemical characterization of a lysosomal protease deficient in classical late infantile neuronal ceroid lipofuscinosis (LINCL) and development of an enzyme-based assay for diagnosis and exclusion of LINCL in human specimens and animal models. *J. Neurochem.* **73**, 700–711.

Taketo, M. M. (1999). Apc gene knockout mice as a model for familial adenomatous polyposis. *Prog. Exp. Tumor Res.* **35**, 109–119.

Tayler, R. M., and Farrow, B. R. H. (1992). Ceroid lipofuscinosis in the border collie dog: Retinal lesions in an animal model of juvenile Batten disease. *Am. J. Human Genet.* **42**, 622–627.

Tyynela, J., Palmer, D. N., Baumann, M., and Haltia, M. (1993). Storage of saposins A and B in infantile neuronal ceroid-lipofuscinoses. *FEBS Lett.* **330**, 8–12.

Tyynela, J., Baumann, M., Hensler, M., Sandhoff, K., and Haltia, M. (1995). Sphingolipid activator proteins (SAPs) are stored together with glycosphingolipids in the infantile neuronal ceroid-lipofuscinosis (INCL). *Am. J. Med. Genet.* **57**, 294–297.

Tyynela, J., Suopanki, J., Baumann, M., and Haltia, M. (1997a). Sphingolipid activator proteins (SAPs) in neuronal ceroid-lipofuscinosis (NCL). *Pediatrics* **28**, 49–52.

Tyynela, J., Suopanki, J., Santavuori, P., Baumann, M., and Haltia, M. (1997b). Variant late infantile ceroid-lipofuscinosis: pathology and biochemistry. *J. Neuropathol. Exp. Neurol.* **56**, 369–375.

Vance, J. E., Stone, S. J., and Faust, J. R. (1997). Abnormailities in mitochondria-associated membranes and phospholipid biosynthetic enzymes in the *mnd/mnd* mouse model of neuronal ceroid lipofuscinosis. *Biochim. Biophys. Acta* **1344**, 286–299.

Vesa, J., Hellsten, E., Verkruyse, L. A., Camp, L. A., Rapola, J., Santavuori, P., Hofmann, S. L., and Peltonen, L. (1995). Mutations in the palmitoyl protein thioesterase gene causing infantile neuronal ceroid lipofuscinosis. *Nature* **376**, 584–587.

Weissenbock, H., and Rossel, C. (1997). Neuronal ceroid-lipofuscinosis in a domestic cat: Clinical, morphological and immunohistochemical findings. *J. Comp. Pathol.* **117**, 17–24.

Willecke, K., Temme, A., Teubner, B., and Ott, T. (1999). Characterization of targeted connexin32-deficient mice: A model for the human Charcot-Marie-Tooth (X-type) inherited disease. *Ann. N.Y. Acad. Sci.* **883**, 302–309.

Wisniewski, K. E., Rapin, I., and Heaney-Kieras, J. (1988). Clinico-pathological variability in the childhood neuronal ceroid-lipofuscinoses and new observations on glycoprotein abnormalities. *Am. J. Med. Genet.* **Suppl. 5**, 27–46.

Wisniewski, K. E., Golabek, A. A., and Kida, E. (1995). Increased urine concentration of subunit c of mitochondrial ATP synthase in neuronal ceroid lipofuscinosis patients. *J. Inherit. Metabol. Dis.* **17**, 205–210.

Wong, F. S., Dittel, B. N., and Janeway, C. A. J. (1999). Transgenes and knockout mutations in animal models of type 1 diabetes and multiple sclerosis. *Immunol. Rev.* **169**, 93–104.

11

Experimental Models of NCL: The Yeast Model

David A. Pearce*
Center for Aging and Developmental Biology
Department of Biochemistry and Biophysics
University of Rochester School of Medicine and Dentistry
Rochester, New York 14642

    I. Introduction
   II. Yeast as a Model for JNCL
  III. What Does Btn1p Do?
  IV. Yeast as a Therapeutic Model for JNCL
   V. A Yeast Model for INCL
      References

I. INTRODUCTION

Genes encoding predicted proteins with high sequence similarity to CLN3 have been identified in mouse, dog, rabbit, nematode, and the yeast *Saccharomyces cerevisiae* (Lee *et al.*, 1996; Shibuya *et al.*, 1997; Katz *et al.*, 1997; Wilson *et al.*, 1994; Pohl and Aljinovic, 1995).
    The yeast *Saccharomyces cerevisiae* is often described as a model eukaryotic organism, due to the many fundamental similarities shared between itself and the cells, albeit differentiated, of higher eukaryotes such as humans. Studying the function of a particular protein in the unicellular yeast provides a very powerful tool for understanding a protein's function. The entire genome of *S. cerevisiae* has

*Address for correspondence: E-mail david_pearce@urmc.rochester.edu been sequenced, and is predicted to encode around 6100 genes, many of whose homologs have been associated to inherited genetic disorders.

II. YEAST AS A MODEL FOR JNCL

Although CLN1, CLN2, CLN3, CLN5, and CLN8 have been identified, only CLN3 has a homolog in the yeast S. cerevisiae. We previously reported that the corresponding yeast gene, denoted *BTN1*, encodes a nonessential protein that is 39% identical and 59% similar to human CLN3 (Pearce and Sherman, 1997); see Figure 11.1. Essentially, by computer analysis human CLN3 and yeast *BTN1* encode the same protein when we factor in the evolutionary divergences that have clearly occurred from yeast to human. To study the function of Btn1p in yeast and therefore to apply our findings to what CLN3 might be doing in the cell, we deleted the *BTN1* gene from yeast, *btn1-Δ*. This provides us with a situation where we have a single-celled organism that is normal, and one that is identical for its whole genome except that we have deleted one specific gene, *BTN1*. We then refer to this as our "Battens disease yeast," as it is normal except that it is lacking the yeast equivalent to CLN3, which is associated to Batten disease. Having created our "Batten disease yeast," we focused our research on deducing whether this protein is involved in mitochondrial function. Our reasoning behind this was simply because at the time many researchers believed that, due to the accumulation of mitochondrial ATP-synthase subunit c in juvenile neuronal ceroid lypofuscinose (JNCL) that CLN3 must be mitochondrial. In summary, we found that *btn1-Δ* strains had no mitochondrial defects. In particular, oxidative phosphorylation or the ability to derive energy in the mitochondria was not affected. Furthermore, we demonstrated that *btn1-Δ* strains had no apparent differences in the degradation of proteins within the mitochondria, and most tellingly, degradation of mitochondrial ATP-synthase subunit c was unaffected (Pearce and Sherman, 1997). We therefore considered that it was unlikely that Btn1p, and consequently CLN3, were indeed mitochondrial proteins. In addition, it is worth noting that *btn1-Δ* strains do not show evidence of accumulation of an autofluorscent pigment.

The value of yeast as a tool for understanding the function of a protein relies heavily on the ease of which we can delete specific genes from the yeast and then assess the effect that the absence of this gene might have. As described above, we first used our *btn1-Δ* strains to look at a specific biochemical attribute, namely, degradation of mitochondrial ATP-synthase subunit c, as this had been implicated in the disease process of Batten disease. Having eliminated any connection between the function of Btn1p with what was essentially the only clue as to what was the function of Btn1p, and therefore CLN3, we needed to uncover an effect that *btn1-Δ* had on the yeast. In other words, we

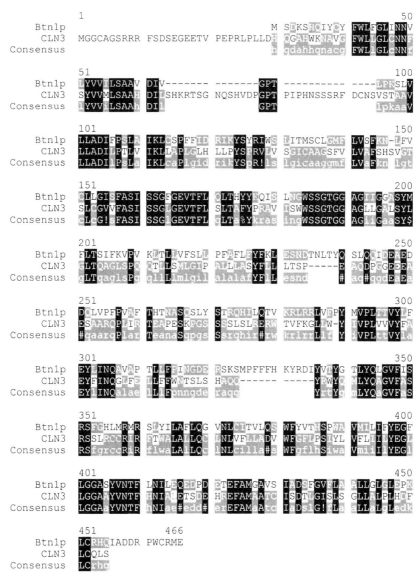

Figure 11.1. Btn1p is homologous to human CLN3. Protein alignment of yeast Btn1p and human CLN3 show that the two proteins show 39% identity and 59% similarity. The protein sequence of Btn1p and human CLN3 is presented followed by a consensus sequence that was generated using MultiAlin (Corpet, 1998). Identical residues are shaded black and similar residues gray. In the consensus sequence, ! represents I or L; $ represents L or M; % represents F or Y; + represents K, R, or Q; # represents B, D, N, Q, or Z; and "represents E, Z, or Q.

needed to uncover a phenotype. We therefore went about comparing the growth of normal and btn1-Δ strains on a variety on media. For example, both strains were grown at elevated temperatures, in the presence of heavy metals or drugs, and growth compared. One such experiment revealed that btn1-Δ, but not normal strains, could grow in the presence of the drug, D-(−)-threo-2-amino-1-[p-nitrophenyl]-1,3-propanediol (denoted ANP), which is derived from chloramphenicol (Pearce and Sherman, 1998). We went on to show that resistance to this drug was dependent on the pH of the growth media. If the pH of media containing ANP was decreased to below pH 6.5, normal yeast gained the ability to grow, whereas if the pH was at pH 6.8 or above, btn1-Δ, but not normal strains, could grow. This difference in essence is a fundamental difference between normal and btn1-Δ yeast strains, our phenotype. To confirm the specificity of this phenotype to the "Batten disease yeast," we introduced a plasmid borne copy of BTN1 into btn1-Δ, and confirmed that the expression of BTN1 restored an inability to grow in the presence of ANP at elevated pH. This specifically demonstrated that our phenotype was due to the presence or absence of BTN1. Furthermore, and most significantly, we expressed plasmid-borne human CLN3 in btn1-Δ and demonstrated that expression of human CLN3 also restored an inability for btn1-Δ to grow in the presence of ANP. This indicates that human CLN3 can substitute the function of Btn1p in btn1-Δ. The fact that CLN3 functions to complement btn1-Δ is strong evidence that CLN3 and Btn1p share the same function within a cell.

Having uncovered a phenotype for btn1-Δ strains, we used two approaches in which we manipulated the sequence of Btn1p to gain further insight into the function of the protein. First we introduced point mutations into BTN1 that corresponded to point mutations in CLN3 that were associated to Batten disease. Second, we fused a fluorescent protein called the green fluorescent protein (GFP) to the amino and carboxyl termini of Btn1p.

The vast majority of individuals with Batten disease have the major 1.02-kb deletion in CLN3, which results in a predicted translation product of the first 153 amino acids rather than the full-length 438 amino acids (International Batten Disease Consortium, 1995). Our studies mentioned above concentrated on a similar situation in yeast whereby we had deleted the entire BTN1 gene. To further the "Battens disease yeast" as a model for studying the disease, we looked at the sequence data on the various point mutations identified in CLN3 that were also associated to the disease as well as the 1.02-kb deletion (International Batten Disease Consortium, 1995). Six missense mutations in CLN3 were reported. The point mutations in CLN3 were leucine → proline at position 101; leucine → proline at position 170; glutamic acid → lysine at position 295; valine → phenylalanine at position 330; arginine → cysteine at position 334; arginine → histidine at position 334. Interestingly, the L101P, L170P, and E295K mutations were judged to present a nonclassical or less severe form of Batten disease, whereas V330F, R334C, and R334H present the classical or severe form of Batten disease. Each of these residues are conserved in the yeast Btn1p, suggesting that they may be important for the correct function of the CLN3 protein. The fact that mutation in these residues leads to Batten disease also suggests that each of these residues is important to the normal functioning of CLN3. We therefore altered the yeast DNA sequence of *BTN1* by *in vitro* mutagenesis such that these mutations now occurred in Btn1p. By expressing each of these mutated forms of Btn1p in a *btn1-Δ* strain, we assayed our phenotype, growth on ANP. Simply put, we measured the ability of the mutant Btn1p to restore the yeast back to the state at which it would be if it contained a normal functional Btn1p. This revealed that the equivalent mutations to L170P and E295K gave virtually no growth on ANP, much like the normal protein, suggesting that Btn1p, and most likely CLN3, with these mutations was almost completely functional. This correlated with the fact that individuals with the L170P and E295K mutations had a less severe course of Batten disease, which in turn suggests that the mutant CLN3 is not completely inactive in its function. Furthermore, the equivalent to the V330F, R334C, and R334H mutations in Btn1p grew well on the ANP, almost like the complete deletion, *btn1-Δ*, suggesting almost complete inactivation of the protein's activity. This again correlated to the disease, as these mutations cause classical or severe Batten disease, which would be expected to be due to inactive CLN3. Overall, we concluded from this that the ANP phenotype can be used as a measure of the functionality of Btn1p and therefore CLN3, which thus far correlates directly to what is known about the severity of the disease. We have recently completed a similar study in which we expressed CLN3 containing the point mutations associated to Batten disease in a *btn1-Δ* strain and confirmed our ANP phenotype to Batten disease severity model using the human protein (Haskell *et al.*, 2000)

The second manipulation of the Btn1p protein sequence simply placed the GFP at the amino and carboxyl termini of the protein. The rationale for doing this was simple. With the presence of the fluorescent tag on the protein, we could visualize the location of this tagged Btn1p under the microscope. Most important, by assessing the growth of a *btn1-Δ* strain expressing these Btn1p-GFP fusions on ANP, we would know if the protein was functional or not. If the Btn1p-GFP fusion is functional, we can assume that the protein is localized to the correct compartment. As it turns out, if the GFP is fused to the carboxy terminus of Btn1p, the protein is indeed functional (Pearce *et al.*, 1999), and localizes to the vacuolar compartment of yeast (Croopnick *et al.*, 1999; Pearce *et al.*, 1999). We can conclude from this that Btn1p is in fact a vacuolar protein, and that CLN3 would localize to the analogous compartment in human cells, the lysosome. Studies of CLN3 have indeed confirmed localization to the lysosome (Jarvela *et al.*, 1998, 1999; Haskell *et al.*, 2000).

III. WHAT DOES Btn1p DO?

The most exciting step in understanding what Btn1p might function as in the yeast cell has been our uncovering the reason why $btn1$-Δ strains are resistant to ANP. Having established that $btn1$-Δ, but not normal strains, are resistance to ANP, we went on to show that resistance to this drug is dependent on the pH of the growth medium. If the pH of media containing ANP was decreased to below pH 6.5, normal yeast gained the ability to grow, whereas if the pH was at pH 6.8 or above, $btn1$-Δ, but not normal strains, could grow (Pearce and Sherman, 1998). This suggested to us that perhaps $btn1$-Δ strains had an inherent ability to decrease the pH of the growth media. It has long been known that as yeast grow, they acidify, or decrease the pH of, the medium. We therefore investigated whether $btn1$-Δ strains have an increased ability to acidify growth media compared to normal. This was performed by simply taking samples of yeast throughout the growth course, resuspending the cells in fresh media, and over time monitoring the pH of this medium using a pH electrode to measure the rate at which the yeast strains acidified the medium. This did in fact demonstrate that $btn1$-Δ strains have an elevated rate of acidification of the medium compared to normal. Curiously, this elevated rate of acidification was demonstrated only in the early phases of growth, and was apparently normalized later in growth. This elevated rate of medium acidification explained quite clearly that the reason that $btn1$-Δ strains are resistant to ANP is that they decrease the pH of the medium containing ANP to the point where ANP is no longer toxic to the cells. The next phase of our research focused on why $btn1$-Δ strains have this elevated rate of acidification.

The yeast plasma membrane H^+-ATPase drives uptake of nutrients into the cell in exchange for protons. It is this activity that results in the growth medium becoming acidic as the yeast grows. We therefore surmised that the activity of the plasma membrane H^+-ATPase may be elevated in $btn1$-Δ strains. To investigate this possibility we isolated the plasma membranes from normal and $btn1$-Δ strains throughout the growth curve and assayed the activity of the plasma membrane H^+-ATPase. This demonstrated that in the early phase of growth $btn1$-Δ strains do have an elevated activity of plasma membrane H^+-ATPase, which correlates directly to our observations on an elevated rate of acidification. Later in growth, activity of plasma membrane H^+-ATPase is apparently normalized, as is the rate of acidification. We therefore concluded that resistance to ANP through an elevated rate of medium acidification in $btn1$-Δ is attributable directly to the elevated activity of plasma membrane H^+-ATPase (Pearce et al., 1999). The next question was why a vacuolar protein such as Btn1p, when deleted, exerts this alteration in the activity of the plasma membrane H^+-ATPase.

It has been widely assumed that the plasma membrane H^+-ATPase must have a general role in the maintenance of intracellular pH homeostasis simply because of its ability to pump protons into the surrounding medium. The vacuole has also been implicated in the maintenance of pH homeostasis by virtue of the fact that it is the major acidic compartment within the cell. With $btn1$-Δ strains having an elevation of proton pumping out of the cell through the plasma membrane H^+-ATPase, it seemed likely that there may be a perturbation in intracellular pH. Using fluorescent dyes that have been specifically characterized to localize to the cytoplasm and vacuole, respectively, and that emit a varying fluorescent signal depending on the pH in this location, we measured cytoplasmic and vacuolar pH in $btn1$-Δ and normal strains. The cytoplasmic-associated dye, fluorescein, carboxyseminaphthorhodafluor-1 (C.SNARF-1), revealed that there was no significant difference in cytoplasmic pH of $btn1$-Δ and normal strains. Both $btn1$-Δ and normal strains had a cytoplasmic pH of 6.8. However, the vacuole-associated dye, 6-carboxyfluorescein diacetate (6-CFDA), revealed that there was a difference in vacuolar pH between $btn1$-Δ and normal strains (Pearce et al., 1999). Specifically, in the early phase of growth, $btn1$-Δ had a more acidic vacuolar pH of 5.8 as compared to 6.2 for normal strains. Again, as with the rate of acidification and the activity of plasma membrane H^+-ATPase, this difference in vacuolar pH became normalized during growth. Vacuolar pH of $btn1$-Δ strains in early growth was 5.8, whereas later in growth vacuolar pH was 6.2. The vacuolar pH of normal strains was consistently 6.2 throughout all phases of growth. It therefore seemed that a decreased vacuolar pH in $btn1$-Δ results in an imbalance in pH homeostasis such that activity, and therefore proton pumping of the plasma membrane H^+-ATPase, is increased. That the elevated rate of acidification, increased activity of plasma membrane H^+-ATPase, and decreased vacuolar pH all returned to normal during growth indicated that $btn1$-Δ strains recognize and correct these abnormalities. To address this, we asked whether there is a fundamental difference between $btn1$-Δ and normal strains by examining and comparing the expression of all yeast genes in $btn1$-Δ and normal strains using a DNA microarray (DeRisi et al., 1997). This revealed that only two yeast genes had significantly different expression between $btn1$-Δ and normal strains. There was no apparent increased gene expression in normal strains compared to $btn1$-Δ, but, $btn1$-Δ had elevated expression of two genes, HSP30 and BTN2.

HSP30 encodes a small heat-shock protein that localizes to the plasma membrane. Furthermore, Hsp30p has been shown to be a stress-induced downregulator of the plasma membrane H^+-ATPase (Piper et al., 1997). This was of immediate interest, as we have shown that $btn1$-Δ has an elevated activity of plasma membrane H^+-ATPase that is returned to normal during growth. Increased expression of HSP30 would therefore decrease the activity of plasma membrane H^+-ATPase. We assessed the expression of HSP30 throughout growth of $btn1$-Δ and normal strains and found that, during growth, expression of HSP30 is increased in $btn1$-Δ at about the point at which plasma membrane H^+-ATPase activity is returned to normal. Interestingly, the level of *HSP30* expression in *btn1-Δ* is maintained at about two times the level of that in normal strains, presumably to continue keeping the plasma membrane H^+-ATPase activity in check.

BTN2 encodes a previously uncharacterized protein of 410 amino acids. Analysis of the expression of *BTN2* throughout the growth curve revealed that its expression was increased at the same time that *HSP30* expression was increased, but as growth continued, *BTN2* expression was six- to sevenfold higher in *btn1-Δ* than in normal strains. This suggests that Btn2p is the major component of correcting for the absence of Btn1p. So what is Btn2p and what does it do? These are questions that we are working on now. A major clue as to the function of Btn2p is that it shares significant homology with the human protein HOOK1. Over a span of 102 amino acids Btn2p is 38% similar to HOOK1, which indicates that while these two proteins are not true homologs, they are most likely related in some way. The corresponding HOOK1 protein from *Drosophila* is a novel protein involved in the endocytosis of transmembrane ligands (Kramer and Phistry, 1996, 1999). *Drosophila* HOOK1 apparently acts as a negative regulator of the fusion of mature multivesicular bodies (MVBs) to the late endosome and lysosome (Sunio et al., 1999). If Btn2p is also involved in a novel endocytic pathway in yeast, an increased expression of *BTN2* in btn1-Δ may compensate the absence of Btn1p. Hypothetically, an increase in Btn2p might decrease fusion of a particular type of vesicle to the yeast vacuole. As these vesicles are most likely acidic, this might contribute indirectly to the slight elevation in the vacuolar pH from 5.8 to 6.2. Clearly, we need to establish exactly what Btn2p does in the yeast model. *BTN2* expression continues to increase through growth of *btn1-Δ*, it would suggest that it might be the primary method of compensating for an absence of Btn1p. An overview of what is occurring in *btn1-Δ* compared to normal strains is presented as a cartoon in Figure 11.2.

So what does Btn1p do? Obviously, there is much work still to be done. Our findings have shown that Btn1p is a vacuolar protein, and that yeast lacking the protein have a defective vacuolar pH, or at least a defective regulation of vacuolar pH. Our most recent studies have confirmed that *btn1-Δ* strains, and strains bearing deletions of either *HSP30* (*hsp30-Δ*) or *BTN2*(*btn2-Δ*), the genes upregulated in *btn1-Δ*, result in a similar phenotype, notably sensitivity for growth in the presence of the weak organic acid, sorbic acid, at low pH (Pearce and Sherman, submitted). This new phenotype ties together the function of Btn1p, Btn2p, and Hsp30p as all being involved in maintaining pH homeostasis. In particular, the nature of sorbic acid suggests that *btn1-Δ*, and also *btn2-Δ* and *hsp30-Δ* strains, have a defect in the ability to grow when presented with a weak acid environmental insult. This strengthens the thought that the absence of Btn1p in yeast and perhaps CLN3 in humans results in a inability to maintain controlled pH homeostasis, particularly at the lysosome.

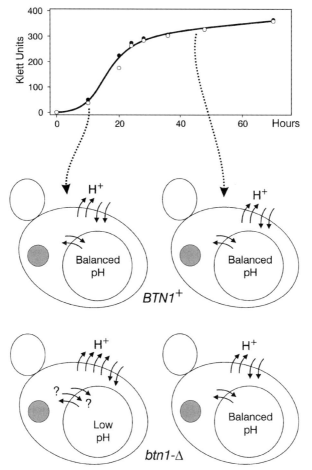

Figure 11.2. pH homeostasis is disturbed in early growth of $btn1$-Δ strains. A schematic representation showing that as normal and $btn1$-Δ strains grow, there is an imbalance of pH within the vacuolar (lysosomal) compartment of $btn1$-Δ strains in the early phase of growth. The $btn1$-Δ strains use the plasma membrane H^+-ATPase to pump more protons into the surrounding medium. As the yeast grow, normal strains maintain a balanced pH environment, while $btn1$-Δ strains using altered gene expression balance the disparity in proton pumping that causes an imbalance in the intracellular pH.

IV. YEAST AS A THERAPEUTIC MODEL FOR JNCL

Having established that an underlying defect in $btn1$-Δ strains is a decreased vacuolar pH early in growth and that yeast use coordinate gene expression to correct this, we addressed the possibility of correcting vacuolar pH with a drug.

Essentially, we wanted to see if we could reverse the effects of lacking Btn1p by treating btn1-Δ with chloroquine.

Chloroquine, a lysosomotropic agent, is widely used as an antimalarial agent, due to its toxicity to Plasmodium falciparum trophozoites (Slater, 1993). Chloroquine accumulates in the acidic food vacuole, causing an increase in pH that is believed to inhibit the mobilization of food reserves during this stage of the parasites development (Goldberg et al., 1990). It is this ability of chloroquine to raise the pH of the acidic vacuolar compartment that prompted us to investigate whether the pH of the vacuole in btn1-Δ yeast strains could be raised. We found that growing btn1-Δ strains in the presence of chloroquine results in the loss of resistance to ANP. During early growth of btn1-Δ strains in the presence of chloroquine, plasma membrane H^+-ATPase activity is decreased and vacuolar pH is increased, changes that result in btn1-Δ resembling normal strains. In essence, we have reversed the biochemical effect of lacking Btn1p. However, as growth continues, plasma membrane H^+-ATPase becomes abnormally low and vacuolar pH becomes abnormally high in both btn1-Δ and normal strains. This essentially can be viewed as a secondary effect, and even other cellular changes may also occur from an abnormally low activity of plasma membrane H^+-ATPase and an abnormally high vacuolar pH. Furthermore, although we have shown that chloroquine does reverse to a degree the increased expression of HSP30 and BTN2 in btn1-Δ, the levels of expression are not returned fully to the levels seen in normal strains. In addition, normal strains are again affected by chloroquine, with HSP30 expression considerably decreased. No doubt chloroquine affects more than just the BTN1-mediated pH homeostasis in yeast, and it would be interesting to examine the effect of chloroquine on the expression of all yeast genes.

It obviously needs to be established whether or not any of the defects found in btn1-Δ strains are truly associated with individuals having Batten disease. Nevertheless, the yeast model is currently a valuable but as yet unproven model to gather information on the pathogenesis of Batten disease.

V. A YEAST MODEL FOR INCL

The previously described "Batten disease yeast model" came about due to the fact that the whole genome of S. cerevisiae was sequenced revealing that this yeast had a protein homologous to CLN3. Recently a second yeast, Schizosaccharomyces pombe, has had its entire genome sequenced. This revealed that S. pombe also contains a homolog to human CLN3. In addition S. pombe contains a protein that is 32% identical and 47% similar to human CLN1 (Figure 11.3), a protein thiolesterase that is associated to infantile NCL (Vesa et al., 1995). The presence of a CLN1 homolog in S. pombe presents the opportunity to study this protein in this organism.

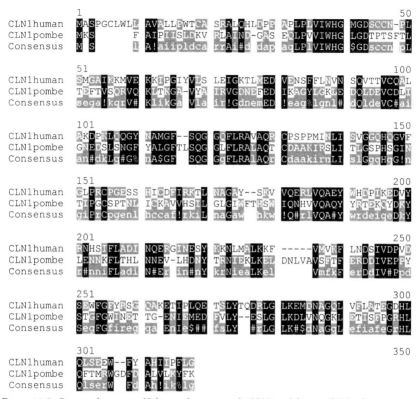

Figure 11.3. Protein alignment of *Schizosacchomyces pombe* CLN1 with human CLN1, demonstrating that the two proteins are 32% identical and 47% similar. The protein sequence of S. pombe CLN1 and human CLN1 is presented followed by a consensus sequence that was generated using MultiAlin (Corpet, 1998). Identical residues are shaded black and similar residues gray. In the consensus sequence, ! represents I or L; $ represents L or M; % represents F or Y; + represents K, R, or Q; # represents B, D, N, Q ,or Z; and "represents E, Z, or Q.

Acknowledgments

I am indebted to Fred Sherman for his support, advice, and for preparing Figure 11.2. The research described herein was supported by National Institutes of Health grant R01 NS36610 and by the JNCL Research Fund.

References

Corpet, F. (1998). Multiple sequence alignment with hierarchical clustering. *Nucleic Acids Res.* **16,** 10881–10890

Croopnick, J. B., Choi, H. C., and Mueller, D. M. (1998). The subcellular location of the yeast *Saccharomyces cerevisiae* homologue of the protein defective in juvenile form of Batten disease. *Biochem. Biophys. Res. Commun.* **250,** 335–341.

DeRisi, J. L., Iyer, V. R., and Brown, P. O. (1997). Exploring the metabolic and genetic control of gene expression on a genomic scale. *Science* **278,** 680–686.

Goldberg, D. E., Slater, A. F. G., Cerami, A., and Henderson, G. B. (1990). Hemoglobin degradation in the human malaria pathogen *Plasmodium falciparum*: A catabolic pathway initiated by a specific aspartic protease. *Proc. Natl. Acad. Sci. (USA)* **87,** 2931–293.

Haskell, R. E., Carr, C. J., Pearce, D. A., Bennett, M. J., and Davidson, B. L. (2000). Batten disease: Evaluation of CLN3 mutations on protein localization and function. *Hum. Mol. Genet.*, in press.

International Batten Disease Consortium (1995). Isolation of a novel gene underlying Batten disease. *Cell* **82,** 949–957.

Jarvela, I., Sainio, M., Rantamaki, T., Olkkonen, V. M., Carpen, O., Peltonen, L., and Jalanko, A. (1998). Biosynthesis and intracellular targeting of the CLN3 protein defective in Batten disease. *Hum. Mol.Genet.* **7,** 85–90.

Jarvela, I., Lehtovirta, M., Tikknen, R., Kyttala, A., and Jalenko, A. (1999). Defective intracellular transport of CLN3 is the molecular basis of Batten disease (JNCL). *Hum.Mol.Genet.* **8,** 1091–1098.

Katz, M. L., *et al.* (1997). Genbank direct submission, accession U92812.

Kramer, H., and Phistry, M. (1996). Mutations in the *Drosophila hook* gene inhibit endocytosis of the boss transmembrane ligand into multivesicular bodies. *J. Cell Biol.* **133,** 1205.

Kramer, H., and Phistry, M. (1999). Genetic analysis of *hook*, a gene required for endocytic trafficking in *Drosophila*. *Genetics* **151,** 675–684.

Lee, R. L., Johnson, K. R., and Lerner, T. J. (1996). Isolation and chromosomal mapping of a mouse homolog of the Batten disease gene CLN3. *Genomics* **35,** 617–619.

Pearce, D. A., and Sherman, F. (1997). *BTN1*, a yeast gene corresponding to the human gene responsible for Batten's disease, is not essential for viability, mitochondrial function, or degradation of mitochondrial ATP synthase. *Yeast* **13,** 691–697.

Pearce, D. A., and Sherman, F. (1998). A yeast model for the study of Batten disease. *Proc. Natl. Acad. Sci. (USA)* **95,** 6915–6918.

Pearce, D. A., Ferea, T., Nosel, S. A., Das, B., and Sherman, F. (1999). Action of Btn1p, the yeast orthologue of the gene mutated in Batten disease. *Nat. Genet.* **22,** 55–58.

Piper, P. W., Ortiz-Calderon, C., Holyoak, C., Coote, P., and Cole, M. (1997). Hsp30, the integral plasma membrane heat shock protein of *Saccharomyces cerevisiae*, is a stress-inducible regulator of plasma membrane H^+-ATPase. *Cell Stress Chaper.* **2,** 12–24.

Pohl, T. M., and Aljinovic, G. (1995). Genbank direct submission, accession Z49335. Shibuya, H., *et al.* (1997). Swiss-Prot direct submission, accession O29611.

Slater, A. F. G. (1993). Chloroquine: Mechanism of drug action and resistance in *Plasmodium falciparum*. *Pharmacol. Ther.* **57,** 203–235.

Sunio, A., Metcalf, A. B., and Kramer, H. (1999). Genetic dissection of endocytic trafficking in *Drosophila* using a horseradish peroxidase-bride of sevenless chimera: *hook* is required for normal maturation of multivesicular endosomes. *Mol. Biol. Cell* **10,** 847–859.

Vesa, J., *et al.* (1995). Mutations in the palmitoyl protein thiolesterase gene causing infantile neuronal ceroid lipofuscinosis. *Nature* **376,** 584–587.

Wilson, R., *et al.* (1994). 2.2 Mb of contiguous nucleotide sequence from chromosome III of *C. elegans*. *Nature* **368,** 32–38.

…
12

Outlook for Future Treatment

Nanbert Zhong*
Department of Human Genetics
New York State Institute for Basic Research
in Developmental Disabilities
Staten Island, New York 10314
and
Department of Neurology
SUNY Health Science Center at Brooklyn
Brooklyn, New York 11203

Krystyna E. Wisniewski
Department of Pathological Neurobiology
New York State Institute for Basic Research
in Developmental Disabilities
Staten Island, New York 10314
and
Department of Neurology
SUNY Health Science Center at Brooklyn
Brooklyn, New York 11203

I. Molecular Cloning for CLN_4, CLN_6, and CLN_7
II. Characterization of Native Substrates for CLN-Encoded Lysosomal Enzymes
III. Proteomic Studies of CLN-Encoded Proteins
IV. Uncovering the Pathogenesis of the NCLs
V. Potential Drugs in Experimental Models May Eventually Lead to Clinical Trials in NCL-Affected Patients
VI. Gene Therapy
References

*Address for correspondence: E-mail: omrddzhong@AOL.com

ABSTRACT

Currently, no treatment is available for neuronal ceroid lipofuscinoses. The progress of human genome project will stimulate molecular cloning of unidentified genes underlying the NCLs, which will lead eventually clinical management and therapies for NCL. Characterizing the native substrate(s) for the palmitoyl-protein thioesterase-1 (PPT1) and tripeptidyl peptidase 1 (TPP1), understanding the protein functions encoded by *CLN* genes, and uncovering the pathological metabolic mechanism for the NCLs are the bases of designing rational treatments for the NCLs. Testing potential therapeutic agents, replacing deficient enzymes, and developing gene therapy will be the major tasks for NCL researchers.

I. MOLECULAR CLONING FOR CLN_4, CLN_6, AND CLN_7

Progress has been made in molecular cloning of CLN_4, CLN_6, and CLN_7. Use of an approach similar to the one for cloning CLN_2 to study adult-onset neuronal ceroid lipofuscinose (ANCL), the recessively inherited adult form of NCL, underlied by CLN_4, suggested that CLN_4 may also encode a lysosomal glycoprotein (Sleat *et al.*, 1998). Locus of CLN_6 has been narrowed recently to 4cM at 15q21-23 by linkage studies (Sharp *et al.*, 1999). Accomplishing molecular cloning of these unidentified genes will be the necessity for designing treatment.

II. CHARACTERIZATION OF NATIVE SUBSTRATES FOR CLN-ENCODED LYSOSOMAL ENZYMES

The future of treatment for NCLs such as infantile NCL (INCL) and late-infantile NCL (LINCL) caused by lysosomal enzymatic deficiency may lie in enzyme replacement therapy. Before such therapy can be developed, (1) the native substrate(s) for the enzyme, (2) the process of enzyme turnover, and (3) the regulation of enzyme expression must be understood. In addition, the *in vitro* expressed and purified lysosomal enzymes must be able to maintain their biological activity for such therapy to be viable.

Two lysosomal enzymes, the palmitoyl-protein thioesterase-1 (PPT1) and tripeptidyl peptidase 1 (TPP1), are known to be involved in INCL and LINCL, respectively. Identification and characterization of their native substrates may lead to better understanding of the pathophysiology of INCL and LINCL; it is clear that different pathogenic mechanisms are involved because the storage material within the lipofuscins (SAPs vs subunit c) is different.

III. PROTEOMIC STUDIES OF *CLN*-ENCODED PROTEINS

The three proteins encoded by CLN_3, CLN_5, and CLN_8 are membranous, but the function of these proteins is unknown. Although the CLN_3 gene has been cloned for 5 years, efforts of generating antibodies for Battenin in the past years were not quite successful. The major difficulty was that Battenin is a highly hydrophobic protein with six (Janes *et al.*, 1996) or eight (Zhong *et al.*, unpublished data) transmembrane domains. Similar difficulties have been noticed for CLN_8-encoded protein (Kaczmarksi *et al.*, unpublished data). Continuing to generate highly sensitive and specific antibodies for *CLN*-encoded proteins will lead to better understanding of protein localization, the process of turnover, and protein–protein interactions for which we (Zhong *et al.*, 2000a) have proposed that these proteins, along with other unidentified proteins, may compose a novel membranous transporting complex. Two such unidentified proteins, BIP_f and BIP_s (*fast* and *slow* Battenin-interactive protein), have been characterized recently as interacting with the CLN_3-encoded proteins Battenin and Battenim (Zhong *et al.*, 1999). Further exploration of this complex may lead to new insights into the functions of *CLN*-encoded proteins.

IV. UNCOVERING THE PATHOGENESIS OF THE NCLs

The NCLs are a group of neurodegenerative disorders sharing common features and phenotypes. Accumulation of mitochondrial subunit c of ATP synthase has been recognized in all NCLs except INCL. This pathological change is a hallmark of the NCLs, but the mechanism whereby similar pathological changes are caused by different genetic deficiencies has been a puzzle to researchers studying NCLs. Recently, the accumulation of subunit c in JNCL has been demonstrated to result from "gain of function" of Battenim, the mutant form of Battenin that derives from 1.02-kb deletion in the CLN_3 gene, through the increased capability of interaction between Battenim and BIP_f (*fast* Battenin-interactive protein), which also interacts with subunit c (Zhong *et al.*, 2000b). This finding may uncover the mechanism of the accumulation of mitochondrial ATPase subunit c in JNCL and lead to a new theory of a "secondary defect" involved in the pathogenesis of the NCLs (Zhong *et al.*, 2000b). Whether proteins encoded by CLN_5 and CLN_8 are also interacting with Battenin or BIP and participating in this secondary mechanism is an interesting question.

This new theory may open a new avenue to pursue for design of a rational therapy.

V. POTENTIAL DRUGS IN EXPERIMENTAL MODELS MAY EVENTUALLY LEAD TO CLINICAL TRIALS IN NCL-AFFECTED PATIENTS

The symptomatic, rather than causal, treatment of giving antioxidants to NCL patients to improve IQ level and to reduce neurological signs and epilepsy was initiated in the 1970s on the basis of the hypothesis that lipofuscins stored in NCLs are due to lipid peroxidation (Santavuori and Moren, 1977; Westermarck and Santavuori, 1984).

Recently, Lamotrigine (LTG) was given to INCL patients to control seizures (Aberg et al., 1997). The severity of seizures in 15/16 (94%) and the frequency of seizures in 14/16 (88%) of patients receiving the treatment were significantly decreased, with no severe side effects. In a subsequent study, LTG was observed to have favorable effects in 23/28 (82%) patients affected with JNCL (Aberg et al., 1999). In light of this, LTG appears to be a valuable drug in treating NCL.

Supplementation with antioxidant polyunsaturated fatty acid in six children affected with JNCL initially showed some potential benefits in slowing clinical progression (Bennett et al., 1994), but the same approach when applied to mnd mice before and after birth does not alter the clinical and pathological course (Bennett et al., 1997).

Experimental evidence obtained from human JNCL and LINCL cases suggested that lipofuscin accumulation is linked to altered carnitine biosynthesis. Dietary supplementation of carnitine in mnd mice, a model of human progressive epilepsy with mental retardation (Ranta et al., 1999), significantly slows the lipofuscin accumulation in the brain neurons and prolongs life span (Katz et al., 1997).

NCLs are progressive neurodegenerative disorders. Applying neurotrophic factors (NTFs) showed promising effects on both the NCL canine neuronal culture model and the mnd mouse model (Dunn et al., 1994; Cooper et al., 1999). This approach may restore "silent" neurons that are positive to response to certain NTFs such as IGF-1 (Cooper et al., 1999). However, it may not be able to control the progressive degeneration of neurons with genetic deficit for CLN-encoded proteins. Whether those "silent" neurons eventually undergo degeneration is an open question.

Bone marrow transplantation (BMT) has been applied in genetic disorders. However, it appears that BMT is not successful in the NCL disorders. This was documented both in native NCL-affected animal models such as English setter dogs (Deeg et al., 1990) and lambs (Westlake et al., 1995), and in human LINCL and JNCL (Lake et al., 1995, 1997), in which BMT did not correct clinical and pathological progression.

It was interesting that chloroquine can reverse phenotypes in btn-1 knockout yeast strain, although not much is known about this model for studying NCL pathogenesis (Pearce et al., 1999). Applying chloroquine into mammalian cells did not reproduce this reversible phenomenon (Golabek et al., personal communication). Whether this chemical can be used as a therapeutic agent in the human NCLs is an open question.

VI. GENE THERAPY

The use of gene therapy in treating the NCL disorders poses a challenge because of the unknown pathogenic mechanism, although genes CLN_{1-3}, CLN_5, and CLN_8 have been cloned. Both in vivo and ex vivo approaches might be applicable. Strategies of gene augmentation to correct "loss of function," therapeutic RNA editing such as ribozymes to target specific transcripts (Phylactou et al., 1998), and allele-specific inhibition of "gain of function" should be our future goals. To approach this goal, NCL animal models, including both native and artificial knockouts, and transgenic technology will play important roles.

In humans, the supplementary gene can be introduced into cells derived from patients (ex vivo approach) or injected into patient tissues (in vivo approach). For central nervous system disorders such as the NCLs, transgenic delivery of the external gene is not quite successful. Direct injection of recombinant virus carrying human erythropoirtin into skeletal muscle provided evidence that systemic gene delivery can be achieved practically (Kessler et al., 1996). Gene transfer by intravenous injection of β-glucuronidase (GUSB) can clear lysosomal storage in a murine model of mucopolysaccharidosis type VII (MPS VII) (Daly et al., 1999), which provided evidence that lysosomal storage disease is treatable.

A major problem researchers faced previously in gene therapy is how to control the gene expression level and the distribution of introduced gene product. Recent progress in studying recombinant virus made it possible for the introduced gene product to be expressed at a high level, not only in neurons but also in astrocytes, under controlled gene copies (Davidson et al., 2000).

Instead of the use of direct injection in the in vivo approach, the challenge of the ex vivo approach for gene therapy for neurological disorders in which the causative gene defect has been identified is to introduce replaced gene into either embryonic stem cells or bone marrow stromal cells. Both types of cells have the potential to be differentiated into neurons (Li et al., 1998; Pereira et al.,

1998). Injection of these cells introduced with vector carrying target gene will allow the potential cells to migrate along the neuron migration pathway and lose immunoactivity (Azizi et al., 1998).

Because INCL and LINCL are caused by lysosomal enzyme deficiency, they should be the best models to start with for gene therapy. Recent progress with knockout mice models provides an advantage for this type of research. The human normal CLN_1 or CLN_2 gene can be engineered into vector to be introduced into the mouse model that does not have normal PPT1 or TPP1 enzyme expression. After the introduction of the CLN_1 or CLN_2 gene, analyses of PPT1 or TPP1 enzymatic activity may guide controlling of enzyme expression *in vivo*.

JNCL, FNCL, and EPMR appear to be caused by membrane protein deficiency. The functions of Battenin (the protein encoded by CLN_3, CLN_5-encoded protein, and CLN_8-encoded protein are unknown, as is the pathogenesis. Whether replacing the mutant protein may reverse the brain damage is an open question.

Our data on Battenin associated with other membranous proteins (Zhong et al., 1999) has led us to hypothesize a novel trans-membranous complex that consists of Battenin, BIP_s, BIP_f, and possibly proteins encoded by CLN_5 and CLN_8. In addition, finding "gain of function" for the Battenim may lead to a new approach for designing a rational therapy for JNCL. In this case, introducing Battenim-specific ribozymes or antisense approach to catalyze Battenim transcripts may be applied.

References

Aberg, L., Heiskala, H., Vanhanen, S. L., Himberg, J. J., Hosking, G., Yuen, A., and Santavuori, P. (1997). Lamotrigine therapy in infantile ceroid lipofuscinosis (INCL). *Neuropediatrics* **28,** 77–79.
Aberg, L., Kirveskari, E., and Santavuori, P. (1999). Lamotrigine therapy in juvenile ceroid lipofuscinosis. *Epilepsia* **40,** 796–799.
Azizi, S. A., Stokes, D., Augelli, B. J., Digirolamo, C., and Prockop, D. J. (1998). Engraftment and migration of human bone marrow stromal cells implanted in the brains of albino rats-similarities to astrocyte grafts. *Proc. Natl. Acad. Sci. (USA)* **95,** 3908–3913.
Bennett, M. J., Gayton, A. R., Rittey, C. D., and Hosking, G. P. (1994). Juvenile neuronal ceroid-lipofuscinosis: Developmental progress after supplementation with polyunsaturated fatty acids. *Dev. Med. Child. Neurol.* **36,** 630–638.
Bennett, M. J., Boriack, R. L., and Birch, D. G. (1997). In-utero and post-delivery supplementation of motor neuron degeneration mutant mice with polyunsaturated fatty acids does not alter the clinical or pathological course. *Neuropediatrics* **28,** 82–84.
Cooper, J. D., Messer, A., Feng, A. K., Chua-Couzens, J., and Mobley, W. C. (1999). Apparent loss and hypertrophy of interneurons in a mouse model of neuronal ceroid lipofuscinosis: Evidence for partial response to insulin-like growth factor-1 treatment. *J. Neurosci.* **19,** 2556–2567.
Daly, T. M., Vogler, C., Levy, B., Haskins, M. E., and Sands, M. S. (1999). Neonatal gene transfer leads to widespread correction of pathology in a murine model of lysosomal storage disease. *Proc. Natl. Acad. Sci. (USA)* **96,** 2296–2300.

12. Outlook for Future Treatment 223

Davidson, B. L., Stein, C. S., Heth, J. A., Martins, I., Kotin, R. M., Derksen, T. A., Zabner, J., Ghods, A., and Chiorini, J. A. (2000). Recombinant adeno-associated virus type 2, 4, and 5 vectors: Transduction of variant cell types and regions in the mammalian central nervous system. *Proc. Natl. Acad. Sci. (USA)* **97**, 3428–3432.

Deeg, H. J., Shulman, H. M., Albrechtsen, D., Graham, T. C., Storb, R., and Koppang, N. (1990). Batten's disease: Failure of allogeneic bone marrow transplantation to arrest disease progression in a canine model. *Clin. Genet.* **37**, 264–270.

Dunn, W. A., Raizada, M. K., Vogt, E. S., and Brown, E. A. (1994). Growth factor-induced growth in primary neuronal cultures of dogs with neuronal ceroid lipofuscinosis. *Int. J. Dev. Neurosci.* **12**, 185–196.

Janes, R. W., Munroe, P. B., Mitchison, H. M., Gardiner, R. M., Mole, S. E., and Wallace, B. A. (1996). A model for Batten disease protein CLN3: Functional implications from homology and mutations. *FEBS Lett.* **399**, 75–77.

Katz, M. L., Rice, L. M., and Gao, C. L. (1997). Dietary carnitine supplements slow disease progression in a putative mouse model for hereditary ceroid-lipofuscinosis. *J. Neurosci. Res.* **50**, 123–132.

Kessler, P. D., Podsakoff, G. M., Chen, X., McQuiston, S. A., Colosis, P. C., Matelis, L. A., Kurtzman, G. J., and Byrne, B. J. (1996). Gene delivery to skeletal muscle results in sustained expression and systemic delivery of a therapeutic protein. *Proc. Natl. Acad. Sci. (USA)* **93**, 14082–14087.

Lake, B. D., Henderson, D. C., Oakhill, A., and Vellodi, A. (1995). Bone marrow transplantation in Batten disease. *Am. J. Med. Genet.* **57**, 369–373.

Lake, B. D., Steward, C. G., Oakhill, A., Wilson, J., and Perham, T. G. (1997). Bone marrow transplantation in late infantile Batten disease and juvenile Batten disease. *Neuropediatrics* **28**, 80–81.

Li, M., Pevny, L., Lovell-Badge, R., and Smith, A. (1998). Generation of purified neural precursors from embryonic stem cells by lineage selection. *Curr. Biol.* **8**, 971–974.

Pearce, D. A., Carr, C. J., Das, B., and Sherman, F. (1999). Phenotypic reversal of the defects in yeast by chloroquine: A model for Batten disease. *Proc. Natl. Acad. Sci. (USA)* **96**, 11341–11345.

Pereira, R. F., O'Hara, M. D., Laptev, A. V., Halford, K. W., Pollard, M. D., Class, R., Simon, D., Livezey, K., and Prockop, D. J. (1998). Marrow stromal cells as a source of progenitor cells for nonhematopoietic tissues in transgenic mice with a phenotype of osteogenesis imperfecta. *Proc. Natl. Acad. Sci. (USA)* **95**, 1142–1147.

Phylactou, L. A., Kilpatrick, M. W., and Wood, M. J. A. (1998). Ribozymes as therapeutic tools for genetic disease. *Hum. Mol. Genet.* **7**, 1649–1653.

Ranta, S., Zhang, Y., Ross, B., Lonka, L., Takkunen, E., Messer, A. M., Sharp, J., Wheeler, R., Kusumi, K., Mole, S., Liu, W., Soares, M. B., de Fatima Bonaldo, M., Hirvasniemi, A., de la Chapelle, A., Gilliam, T. C., and Lehesjoki, A. (1999). The neuronal ceroid lipofuscinoses in human EPMR and *mnd* mutant mice are associated with mutations in *CLN8*. *Nat. Genet.* **23**, 233–2236.

Santavuori, P., and Moren, R. (1977). Experience of antioxidant treatment in neuronal ceroid-lipofuscinosis of spielmeyer-sjogren type. *Neuropadiatrics* **8**, 333–344.

Sharp, J. D., Wheeler, R. B., Lake, B. D., Fox, M., Gardiner, R. M., and Williams, R. E. (1999). Genetic and physical mapping of the CLN6 gene on chromosome 15q21-23. *Mol. Genet. Metab.* **66**, 329–331.

Sleat, D. E., Sohar, I., Pullarkat, P. S., Lobel, P., and Pullarkat, R. K. (1998). Specific alterations in levels of mannose 6-phosphoryled glycoproteins in different neuronal ceroid lipofuscinoses. *Biochem. J.* **334**, 547–551.

Westermarck, T., and Santavuori, P. (1984). Principle of antioxidant therapy in neuronal ceroid lipofuscinosis. *Med. Biol.* **62**, 148–151.

Westlake, V. J., Jolly, R. D., Jones, B. R., Mellor, D. J., Machon, R., Zanjani, E. D., and Krivit, W. Hematopoietic cell transplantation in fetal lambs with ceroid-lipofuscinosis. *Am. J. Med. Genet.* **57**, 365–368.

Zhong, N., Ju, W., Moroziewicz, D., Jurkiewez, A., Wisniewski, K., and Brown, W. T. (1999). Molecular pathogenic studies of Batten disease: Identification and characterization of Battenin-interactive proteins. *J. Mol. Diag.* **1,** G23 (Abstr.).

Zhong, N., Ju, W., Moroziewicz, D., Jurkewez, A., Wisniewski, K., and Brown, W. T. (2000a). CLN-encoded proteins do not interact with each other. *Neurogenetics* **86,** in press.

Zhong, N., Ju, W., Moroziewicz, D., Jurkewez, A., Wisniewski, K., and Brown, W. T. (2000b). Proteins BIPs and BIPf interact with Battenin/Battenim and may play central roles in the pathegenesis of NCL. *The 8th International Congress on Neuronal Ceroid Lipofuscinoses*, Sept. 20–24, 2000, Oxford, UK (Abstr.).

Appendix: Batten Support Groups

UNITED STATES OF AMERICA

Batten Disease Support and Research Association (BDSRA)
(includes USA, Canada, Australia, and New Zealand)
2600 Parsons Avenue
Columbus, OH 43207
Contact: Mr. Lance Johnston, Executive Director
Phone: (800) 448-4570 and (740) 927-4298
E-mail: bdsra1@bdsra.org
Web: http://bdsra.org

Children's Brain Diseases Foundation
350 Parnassus Avenue, Suite 900
San Francisco, CA 94117
Contact: Dr. Alfred Rider
Phone: (415) 566-5402
Fax: (415) 863-3452

Batten Disease Registry
NYS Institute for Basic Research in Developmental Disabilities
1050 Forest Hill Road
Staten Island, NY 10314
Contact: Prof. Krystyna E. Wisniewski
Department of Pathological Neurobiology
Phone: (718) 494-5202
Fax: (718) 982-6346
E-mail: BATTENKW@AOL.COM

National Organization for Rare Disorders, Inc. (NORD)
P.O. Box 8923
New Fairfield, CT 06812-8923
Phone: (203) 746-6518, Toll free: (800) 999-6673
Fax: (203) 746-6481
E-mail: orphan@nord-rdb.com *or* orphan@rarediseases.org
Web: http://www.rarediseases.org/*

State and regional chapters of BDSRA

Alabama

3560 Alabama Hwy 155
Jemison, AL 35085
Contact: Becky Lucas
Phone: (205) 668-0812

Delaware Valley

25 Aberdeen Drive
Sicklerville, NJ 08081
Contact: Gregg Froio
Phone: (609) 435-0212

Metro New York/New Jersey

21 Glenda Drive
Deerpark, NJ 11729
Contact: Eric Faret
Phone: (631) 586-4315

Northern California

1721 Mt. Vernon Drive
San Jose, CA 95125
Contact: Liz Aurelio
Phone: (408) 448-2530
E-mail: dlaurelio@earthlink.net

3567 Mignon Street
Sacramento, CA 95826
Contact: Nancy Peterson
Phone: (916) 362-9783

Southern California

5065 Frink Ave.
San Diego, CA 92117
Contact: Shannon Killebrew-Ross
Phone: (619) 276-4441
E-mail: luv2roof@aol.com

Florida

1625 28 Avenue North
St. Petersburg, FL 33713
Contact: Karen Upchurch
Phone: (813) 894-4318
E-mail: kupchurch@prodigy.net

Gulf Coast

23211 Tony Wallace Road
Robertsdale, AL 36567
Contact: Michele Kingston
Phone: (334) 947-6254

Louisiana

806 N. Ferrylake Court
Slidell, LA 70461
Contact: Jim Little
Phone: (504) 781-5260
E-mail: jamesflittle_70461@yahoo.com

Michigan

16022 Blue Skies
Livonia, MI 48154
Contact: Linda Houghby
Phone: (734) 591-3062
E-mail: daveh@e_mail.com

Mid-Atlantic

(MD, DE, DC, NJ, WV, East. PA, and North VA)
13818 Greenfield Ave.
P.O. Box 283
Maugansville, MD 21767-0283
Contact: Vicky Lumm
Phone: (301) 790-1417

Midwest

(IL and WI)
13560 Mohawk Lane
Orland Park, IL 60462
Contact: Barbara McDonough
Phone: (708) 460-3054
E-mail: barblarmcd@aol.com

Minnesota

5376 Cottage Avenue
White Bear Twp, MN 55110
Contact: Joni Metcalf
Phone: (651) 429-3871
E-mail: joni_metcalf@hotmail.com

Mississippi

7825 Frank Snell Road
Pascagoula, MS 39581
Contact: Paula Stampley
Phone: (228) 588-2067

New England

(CT, RI, MA, VT, NH, ME)
11 Ammonoosuc St., Apt. 1A
Woodsville, NH 03785
Contact: Ann Salladin
Phone: (603) 747-4191

New York (upstate)

31 Huntington Parkway
Hamlin, NY 14461
Contact: Kristin Coon
Phone: (716) 964-8734

Ohio

3038 Ramona Ave.
Cincinnati, OH 45211
Contact: Bob Wilhelm
Phone: (513) 481-6149
E-mail: bobwilhelm@aol.com

Pacific Northwest

(WA, OR, ID)
1401 15th Street
Anna Cortes, WA 98221
Contact: Mark Vance
Phone: (360) 299-3027
E-mail:73122.2666@compuserve.com

Southeast

(NC, SC, VA)
5359 Deep River Road
Sanford, NC 27330
Contact: Curtis Lowther
Phone: (919) 774-1933

Tennessee

520 Green Meadows Road
Rockwood, TN 37854
Contact: Ethel Flanagan
Phone: (615) 354-3619

Texas

1535 Jerry Maschek St.
West Texas, TX 76691
Contact: Michele Neal
Phone: (254) 826-4651
E-mail: bdsra@flash.net

Web site for current information on each chapter

www.bdsra.org

CANADIAN CHAPTER

12 Bell Street
Regina, Saskatchewan S4S 4B7
Contact: Bev Maxim
Phone: (306) 789-9047
E-mail: gmaxim@cableregina.com

Battens list on the Web

(particularly concerned with caring for a child with Batten disease)
1556 Union Avenue, Box 189
Ruthven, Ontario N0P1G0, Canada
Contact: Eadie Mastronardi
Phone: (519) 322-2987
Fax: (519) 326-1417
E-mail: britjar@wincom.net

EUROPEAN SUPPORT GROUPS

The Czech Republic

The Prague NCL Group
Institute for Inherited Metabolic Disorders
1st Faculty of Medicine and General University Hospital
Unemocnice 5, CZ-12853
Prague 2, Czech Republic
Contact: Milan Elleder, M.D., Ph.D.
Phone: +420 2 295 202
Fax: +420 2 2491 6306
E-mail: melleder@cesnet.cz

Denmark

Dansk Spielmeyer-Vogt Forening
Korsevaenget 58
DK-4100 Ringsted, Denmark
Chairman: Ole Tornvig-Christensen
Phone and Fax: +45 57 61 10 15
E-mail: otc@olicom.dk

Finland

Finnish INCL-Association
via Finnish Association on Mental Retardation
Valokkiviita 5
FIN-60150 Seinäjoki, Finland
Special worker: Pirjo Saari for INCL
Phone: +358 6 4192701
Fax: +358 6 4192705
E-mail: pirjo.saari@famr.fi

Spielmeyer-Janský Association
Via Federation for the Visually Handicapped
Mäkelänkatu 50
FIN-00510 Helsinki, Finland
Special worker: Maria-Liisa Punkari for LINCL and JNCL
Phone: +358 9 3960 4329
Fax: +358 9 3960 4200

Family Federation of Finland
(for genetic counseling only)
P.O. Box 849
FIN-00101 Helsinki, Finland
Phone: +358 9 616 22246
Fax: +358 9 645 018

France

Association Vaincre les Malades Lysosomales
9, place du 19 mars 1962—Parc Affaires 2000
F-91035 Evry CEDEX, France
Phone: +33 1 60 91 75 00
Fax: +33 1 69 36 93 50
E-mail: VML@provnet.fr
Web: http://www.provnet.fr/VML/fr/accueil.htm

Laboratoire de Génétique (Professeur Livia Poenaru)
Faculté de Médécine Cochin Port-Royal
24, rue du Faubourg Saint Jacques
F-75014 Paris, France
Contact: Dr. Catherine Caillaud, M.D., Ph.D.
Phone: +33 1 44 41 24 02 (direct), +33 1 42 34 15 74 (secretary)
Fax: +33 1 44 41 24 46
E-mail: caillaud@cochin.inserm.fr

Germany

NCL Gruppe Deutschland e.V.
Vierkaten 32b
D-21269 Neu-Wülmsdorf, Germany
Phone and Fax: +49 40 700 7521

Italy

COMETA/A.S.M.M.E.
Via Garibaldi 8, Legnaro
I-Padova 35020, Italy
Contact: Mrs. Marzenta, president of the group
Dr. Alberto Burlina, medical support
Phone and Fax: +39 049 9903303

The Netherlands

De Nederlandse Vereniging Batten-Spielmeyer-Vogt (BSV)
President: Mr. K. Kovacsevits
De Ruijterstraat 11
NL-2851 Haastrecht, The Netherlands
Phone: +31 1825 01967
Fax: +31 1825 22822
E-mail: 106333.55@compuserve.com

Secretary: Mrs. Anna Agricola
De Rijlst 38
8521 NJ St Nicolaasga
Friesland, Holland, Nederland
Phone: +31 513 432344
E-mail: 106333.55@compuserve.com

Mw Drs. Irene Hofman (medical consultant)
C/o Bartimeushage, Postbox 87
NL-3940 AB Doorn, The Netherlands
Phone and Fax: +31 317 615191
or
Bergweg 30, NL-3911 VB Rhenen
The Netherlands
Phone and Fax: +31 317 615191
E-mail: 106333.55@compuserve.com

Norway

Norsk Spielmeyer—Vogt forening (NCL)
P.O. Box 55
N-4262 Avaldsnes, Norway
Chairman: Mr. Reidar Dombestein
Phone and Fax: +47 88004474
E-mail: paulande@online.no

Sweden

Parents association for Spielmeyer—Vogt is part of the Swedish association for the visually impaired (SRF Synskadades Riksförbund)
Sandsborgsvägen 52
S-122 88 Enskede, Sweden
Contact: Eva Setreus
Phone: +46 8 39 90 00
Fax: +46 8 39 93 22

Pedagogical, psychological and social support for JNCL families:
Spielmeyer—Vogt support person (consultant)
Contact: Hjördis Gustavsson
Ekeskolan, Box 9024
S-700 09 Örebro, Sweden
Phone: +46 19-676 21 00
Fax: +46 19 676 22 00

Coordinator for medical support board:
Contact: Paul Uvebrant
Sahlgrenska University Hospital Östra
S-416 85 Göteborg, Sweden
Phone: +46 31 343 47 31
Fax: +46 31 25 79 60
E-mail: paul.uvebrant@sahlgrenska.se

Switzerland

Retinitis pigmentosa—Vereinigung Schweiz
Langstrasse 120
CH-8004 Zürich, Switzerland
Contact: Mrs. Ch. Fasser
Phone: +44 1 2911872
Fax: +41 1 2911850

United Kingdom

Batten Support and Research Trust, Reg. Charity No. 107095
Contact. Mr. R J Ford, 83 Weymouth Bay Avenue
Weymouth, Dorset DT3 5AD, England, U.K.

Batten Disease Family Association, c/o Heather House
Heather Drive, Tadley
Hampshire RG26 4QR, England, U.K.
Manager: Sarah Kenrick
Heather House offers services to young adults with a visual impairment and a degenerative illness, especially juvenile NCL. Provision will include a nursing and social care service; an activity and resource center; family support; outreach services; research. Contact the manager, or placement coordinator (below) for further information, to pursue a referral, or to arrange a visit.

SeeAbility
56-66 Highlands Road
Leatherhead, Surrey KT22 8NR, England, U.K.
Placement Coordinator for Heather House: Sue Ogden
Phone: +44 1372 373086

Research Trust for Metabolic Diseases in Children (RTMDC)
Golden Gates Lodge, Weston Road
Crewe, Cheshire CW2 5XN, England, U.K.
Contact: Mrs. Lesley Green, Director of Support Services
Phone: +44 1270 250221
Fax: +44 1270 250224
Language: French and German available

Society for the Study of Inborn Errors of Metabolism (SSIEM)
Hon Secretary: Dr. J. H. Walter
Willinck Biochemical Genetics Unit
Royal Manchester Children's Hospital
Pendlebury, Manchester M27 4HA, England, U.K.
Phone: +44 161 727 2137/8
Fax: +44 161 727 2137
Web: http://www.ssiem.org.uk/

ELSEWHERE

Australia

The Australian Chapter of the Batten Disease Support and Research Association
Contact: Harry Partridge
Level 4, 1 Chandos St
St. Leonards, NSW 2065, Australia
Phone and Fax: +61 2 9460 9000
E-mail: ppstruct@zemail.com.au

Costa Rica

Association Pro Niños Con Enfermedades Progresivas (Batten), APRONEP
Contact: Mr. Yamileth Chávez Soto, Education Committee Coordinator
75 metros al sur de la Fábrica Neón Nieto
San Juan de Tibás, San José, Costa Rica
Phone and Fax: +506 236 96 20

New Zealand

New Zealand Affiliate of BDSRA
P.O. Box 232, Old Station Road
Ohakune, New Zealand
Contact: Kerry Parkes
Phone and Fax: +011-06-385 8103

South Africa

Contact: Mr. Marius VanZyl
Roy Campbell 46,
Sasolburg 9570
Orange Free State
So. Africa
Phone: +011-27 16 976 5294

Venezuela

Fundacion Thairy Rojas Funtharo (Thairy Rojas Foundation)
Contact: Mr. Elimar Rojas or Mrs. Miriam Rojas
Avenida 14, Casa No. 27, Sector Carabobo, Urb. Tamare.
Ciudad Ojeda, Zulia, Venezuela ZP-4017
Phone: +58 65 405740
E-mail: funtharo@hotmail.com *or* fliarojas@hotmail.com *or* elimar.rojas@eudoramail.com

Contact: Dr. Joaquin Peña
Facultad de Medicina
Departmento de Pediatría
Hospital Universitario
Maracaibo. Venezuela
Urb. Canaima, Calle 42, No. 15-18
Maracaibo, Zulia, Venezuela
Phone: +58 61 494408
Fax: +58 61 492186
E-mail: jokar1@telcel.net.ve

Subject Index

A

Adult neuronal ceroid lipofuscinosis, 8, 24
AF, see Amniotic fluid
Amniotic fluid, NCL testing, 152–153
ANCL, see Adult neuronal ceroid lipofuscinosis
Animal models, for human disorders
  for ceroid lipofuscinoses, 184
    CLN8 homolog, 137–138
    English setter dog, 186–188
    JNCL, 117–118, 195–197
    LINCL, 197–198
    mnd mouse, 194
    mouse, 170–173
    nclf mouse, 195
    South Hampshire sheep, 188, 193
    Tibetan terrier dog, 193–194
H^+-ATPase
  and Btn1p, 210–212
  in lysosome pH, 53

B

Batten disease, see Neuronal ceroid lipofuscinoses
Batten support groups, 225–236
Behavior
  in Finnish LINCL, 126
  in Northern epilepsy, 127
Blood, peripheral, in NCLs, 143–144
Brain
  CLN2 neurons, 45
  CLN3 subject, 49–50
BTN1
  JNCL yeast model, 206–209, 213–214
  protein function, 210–212

C

Carrier status, NCLs, 151–152, 155, 162–164
Central nervous system, NCL mouse models, 171–173
Ceroid lipofuscinoses, animal models
  English setter dog, 186–188
  mnd mouse, 194
  nclf mouse, 195
  necessity, 184
  South Hampshire sheep, 188, 193
  Tibetan terrier dog, 193–194
Chorionic villus samples, NCL testing, 152–153
Chromosome 16, CLN3 mapping, 108
Clinical data
  Finnish LINCL, 125–126
  Northern epilepsy, 126–127
Clinopathology, NCL cases, 8
CLN
  encoded lysosomal enzymes, 218–219
  proteomics, 219
CLN1
  characterization, 185
  definitive diagnosis, 14–15
  electrophysiology studies, 14
  EM studies, 14
  linkage disequilibrium mapping, 71
  mutations, 39, 154
  PCR amplification, 149
  PPT1 issues, 39–42
  PPT1 role, 96–97
  PPT deficiency, 81–82
  PPT mutations, 82–86
  SAPs, 40–42
  symptoms, 14
  in variant juvenile NCL, 15–16
CLN2
  biochemical assay, 18–19
  brain neurons, 45
  characterization, 185
  defective protein, 46–47
  EEG, 18
  features, 42
  gene scanning, 151
  incidence, 16–18
  lysosomal storage, 42–45

237

mutation testing, 151, 154
pathogenesis, 46
PCR amplification, 149
TPP1, 18–19, 97–98
CLN3
  amino acid sequence, 51
  animal models, 117
  brain tissue, 49–50
  categories, 21
  characterization, 185–186
  chromosomal mapping, 108
  deletion testing, 144–147
  encoded protein, 116–117
  English setter dog model, 187–188
  exon trapping, 113
  function, 53
  gene scan, 148
  glycosylation, 51
  heterozygosity, 22
  homozygosity, 22
  identification, 50, 98–99
  incidence, 19–21
  localization, 109–111
  mouse knockout models, 195–197
  mouse models, 171–172
  mutations, 52, 115–116, 147–149, 154
  neuronal death, 54
  physical mapping, 111–112
  sequence homology, 52–53
  Southern hybridization, 149
  structural modeling, 50
  as subunit 9 gene, 109
  tissue expression, 116
  yeast model, 206–209
CLN4
  electrophysiological studies, 24
  gene cloning, 218
  genetic defects, 58–59
  symptoms, 24
CLN5
  characteristics, 24–25
  computer predictions, 134
  gene defects, 54
  gene isolation, 132
  mutations, 133–134
  tissue expression, 134
CLN6
  characteristics, 25
  gene cloning, 218
  genetic defects, 56–57

CLN7
  characteristics, 25, 29
  gene cloning, 218
  genetic defects, 57–58
CLN8
  characteristics, 29
  computer predictions, 135
  gene isolation, 134
  genetic defects, 54–56
  mouse homolog, 137–138
  mutation, 134–135
  tissue expression, 135
CNS, *see* Central nervous system
Computer modeling
  CLN5, 134
  CLN8, 135
CVS, *see* Chorionic villus samples

D

Data, clinical
  Finnish LINCL, 125–126
  Northern epilepsy, 126–127
Degeneration, neurons, NTF role, 177–178
Deletions
  CLN3, 144–147, 149
  JNCL mutation, 113–114
Dementia
  in Finnish LINCL, 125–126
  in Northern epilepsy, 126
DNA
  CLN3 deletion, 145
  JNCL CLN3 cDNA, 113
  NCLs, genetic testing, 143–144
  PPT cDNA, 79–80
Drugs, NCL treatment, 220–221

E

EEG, *see* Electroencephalography
Electroencephalography
  for classic late-infantile NCL, 16–18
  Finnish LINCL, 127
  Northern epilepsy, 127–128
Electron microscopy
  INCL, 14
  NCLs, 3
  variant late-infantile NCL with GROD, 15
EM, *see* Electron microscopy
English setter dog, as human disorder model, 186–188

Subject Index

Enzymes, lysosomal
  CLN-encoded, substrates, 218–219
  oligosaccharide role, 76–77
Enzymology, PPT, 73–74
Exons, for JNCL *CLN3*, 113

F

Fatty acids, PPT hydrolysis, 73
Finnish late-infantile neuronal ceroid
    lipofuscinosis
  biochemistry, 130
  causative mutation, 82
  characteristics, 24–25
  clinical data, 125–126
  CLN1 linkage disequilibrium mapping, 71
  cytochemistry, 130
  diagnosis, 135–136
  genetic defects, 54
  histology, 128–129
  neurophysiology, 127
  neuroradiology, 128
  treatment, 137
  ultrastructure, 129–130

G

Genes
  BTN1, 206–209
  btn1-Δ, 213–214
  CLN1, 14–16, 37–42, 71, 149, 154
  CLN2, 16–19, 42–49, 149–151, 154
  CLN3, 19–23, 49–54, 108–113, 115–117,
      144–149, 154, 171–172, 187–188,
      195–197
  CLN4, 24, 58–59, 218
  CLN5, 24–25, 54, 132–134
  CLN6, 25, 56–57, 218
  CLN7, 25, 29, 57–58, 218
  CLN8, 29, 54–56, 134–135, 137–138
  NCL-associated genes, 10–12
  PPT, 79–80
  subunit 9, 109
  TPP1, 18–19
Gene therapy, NCL disorders, 221–222
Genetic counseling, for NCLs, 161–162,
    164–165
Genetic defects
  NCLs, 37, 54–56, 96–99, 160–161
  NCLs without, 56–59

Genetic testing
  for INCL, 149–151
  for JNCL, 144–149
  for LINCL, 149–151
  for NCLs, 143–144, 154
  for prenatal NCLs, 153
Glycosylation, CLN3p, 51
Granular osmiophilic deposits
  in INCL, 14–15
  in JNCL, 15–16
  in LINCL, 15
  in NCLs, 3
GROD, *see* Granular osmiophilic deposits

H

Histology
  Finnish LINCL, 128–129
  Northern epilepsy, 131
Human disorders, animal models,
    184–186
  English setter dog, 186–188
  mnd mouse, 194
  nclf mouse, 195
  South Hampshire sheep, 188, 193
  Tibetan terrier dog, 193–194

I

INCL, *see* Infantile neuronal ceroid
    lipofuscinosis
Infantile neuronal ceroid lipofuscinosis
  characterization, 185
  definitive diagnosis, 14–15
  electrophysiology studies, 14
  EM studies, 14
  mutation testing, 39, 149–151, 154
  PPT1 deficiency, 39–42
  PPT role, 78–79
  SAPs, 40–42
  symptoms, 3, 14
  treatment, 86–88
  yeast model, 214
Insulin-like growth factor-1,
    178–179

J

JNCL, *see* Juvenile neuronal ceroid
    lipofuscinosis

Subject Index

Juvenile neuronal ceroid lipofuscinosis
  amino acid sequence, 51
  animal models, 117–118
  brain tissue, 49–50
  categories, 21
  characterization, 185–186
  common mutation, 113–114
  gene deletion, 144–147
  gene exon trapping, 113
  gene identification, 50
  gene localization, 109–111
  gene mapping, 108
  gene mutations, 52, 115–116, 147–149, 154
  with GROD, 15–16
  heterozygosity, 22
  homozygosity, 22
  incidence, 19–21
  mouse gene knockout models, 195–197
  neuronal death, 54
  PPT mutations, 83
  protein function, 53
  protein sequence homology, 52–53
  protein structural modeling, 50
  subunit 9 gene, 109
  symptom onset, 3
  tissue expression, 116
  yeast model, 206–209, 213–214

K

Kufs' disease, genetic defects, 58

L

Late-infantile–early-juvenile neuronal ceroid lipofuscinosis, 25, 56–57
Late-infantile neuronal ceroid lipofuscinosis
  biochemical assay, 18–19
  brain neurons, 45
  characterization, 185
  defective protein, 46–47
  EEG, 18
  features, 42
  gene scan, 151
  with GROD, 15
  incidence, 16–18
  lysosomal storage, 42–45
  molecular testing, 149–151
  mouse gene knockout models, 197–198
  mutation testing, 154
  pathogenesis, 46
  symptom onset, 3
  TPP1 analysis, 18–19
Late-onset neuronal ceroid lipofuscinosis, 83, 86
LINCL, *see* Late-infantile neuronal ceroid lipofuscinosis
Linkage disequilibrium mapping, *CLN1*, 71
Lysosomal-associated membrane proteins, 52
Lysosomal enzymes
  CLN-encoded, substrates, 218–219
  oligosaccharide role, 76–77
Lysosomal storage
  in CLN1, 40–41
  in CLN2, 42–45
  NCLs as disease
    and mannose 6-phosphorylated glycoproteins, 101
    and mitochondrial ATP synthase subunit c, 100
    and oligosaccharyl diphosphodolichol, 99–100
    and saposins, 101
Lysosomes
  pH from V-ATPase, 53
  targeting, PPT, 75–77

M

Mannose 6-phosphorylated glycoproteins, 101
Mapping
  chromosome 16, *CLN3*, 108
  CLN3, 111–112
  linkage disequilibrium, *CLN1*, 71
  physical, *CLN3*, 111–112
Mitochondrial ATP synthase subunit c
  in CLN2 lysosomal storage, 42–45
  in CLN2 pathogenesis, 45–46
  in CLN3, 50
  in NCL, 100
  and progressive neuronal degradation, 48
  as TTPI substrate, 47–48
Models
  animal, *see* Animal models
  CLN5, 134
  CLN8, 135
  potential NCL drugs, 220–221
  structural, for CLN3p, 50
  yeast, for NCLs, 206–209, 213–214

Subject Index

Molecular genetics
  CLN5, 132–134
  CLN8, 134–135
  PPT deficiency, 81–82
  PPT mutations, 82–86
Molecular screening, NCL carrier status, 151–152, 155
Mouse model
  CLN8 homolog, 137–138
  CNS characterization, 171–173
  gene knockout models, 195–198
  mnd mouse, 194
  nclf mouse, 195
  NCLs, 170–171
Mutations
  CLN3, 52, 115–116
  CLN5, 133–134
  CLN8, 134–135
  INCL, 39, 149–151
  JNCL, 113–114, 147–149
  LINCL, 149–151
  NCL, 10–12, 153–154
  PPT, 82–86
  TPP1, 47

N

NCLs, *see* Neuronal ceroid lipofuscinoses
Neurodegenerative disorders, NTFs as therapeutic agents, 173–176
Neuronal ceroid lipofuscinoses
  ANCL, 8, 24
  animal models, 184–186
  associated genes, 10–12
  carrier status, 151–152, 155, 162–164
  clinopathology, 8
  definition, 3
  diagnosis, 10, 161
  Finnish LINCL, 24–25, 54, 71, 82, 125–130, 135–137
  gene therapy, 221–222
  genetic counseling, 161–162
  genetic defects, 37, 54–56
  genetics, 160–161
  genetic testing, 143–144, 155
  groups, 36
  IGF-1 treatment, 178–179
  INCL, *see* Infantile neuronal ceroid lipofuscinosis

inclusion bodies, 3
inheritance, 160
JNCL, *see* Juvenile neuronal ceroid lipofuscinosis
late-onset NCL, 83, 86
LINCL, *see* Late-infantile neuronal ceroid lipofuscinosis
LINCL–JNCL, 25, 56–57
mannose 6-phosphorylated glycoproteins, 101
mitochondrial ATP synthase subunit c, 100
mouse models, 170–173
NTFs as therapeutic agents, 179
oligosaccharyl diphosphodolichol, 99–100
pathogenesis, 219–220
potential drug models, 220–221
prenatal diagnosis, 153–154
prenatal testing, 152–153
reproductive options, 164–165
saposins, 101
Turkish LINCL, 25, 29, 57–58
without genetic defects, 56–59
Neurons
  CLN2 brain, 45
  death in CLN3, 54
  NTF role, 177–178
  progressive degradation, 48
Neurophysiology
  Finnish LINCL, 127
  Northern epilepsy, 127–128
Neuroradiology
  Finnish LINCL, 128
  Northern epilepsy, 128
Neurotrophic factors
  expression and actions, 176–177
  signaling failure, 177–178
  as therapeutic agents, 173–176, 179
Northern epilepsy
  biochemistry, 132
  characteristics, 29
  clinical data, 126–127
  cytochemistry, 132
  diagnosis, 136–137
  genetic defects, 54–56
  histology, 131
  neurophysiology, 127–128
  neuroradiology, 128
  treatment, 137
  ultrastructure, 131–132
NTFs, *see* Neurotrophic factors

O

Oligosaccharides, in lysosomal enzymes, 76–77
Oligosaccharyl diphosphodilichol, in NCL, 99–100
Ophthalmology
  in Finnish LINCL, 126
  in Northern epilepsy, 127

P

Palmitoyl-protein thioesterase
  cDNA, 79–80
  deficiency
    laboratory diagnosis, 86
    mutations, 82–86
    as recessive genetic trait, 81–82
    treatment, 86–88
  enzymology, 73–74
  fine mapping, 71
  in lysosomal catabolism, 72
  lysosomal targeting, 75–77
  physiological role, 77–79
  posttranslational processing, 75–77
  PPT gene span, 79–80
Palmitoyl-protein thioesterase-1
  in CLN1, 96–97
  in INCL, 39–42
  in JNCL with GROD, 16
Palmitoyl-protein thioesterase-2, 79
Pathogenesis
  CLN2, 46
  NCLs, 219–220
Peripheral blood, in NCLs, 143–144
Physical mapping, *CLN3*, 111–112
Polymerase chain reaction
  CLN1, 149
  CLN2, 149
  CLN3, 144–145, 148
Polymorphisms, NCL-associated genes, 10–12
Posttranslational processing, PPT, 75–77
PPT, *see* Palmitoyl-protein thioesterase
PPT1, *see* Palmitoyl-protein thioesterase-1
PPT2, *see* Palmitoyl-protein thioesterase-2
Prenatal molecular testing, for NCL, 152–153
Progressive neuronal degradation, 48
Prosaposin, 41–42
Proteolytic processing, and PPT, 77
Proteomics, CLN-encoded proteins, 219

S

Saposins
  in CLN1, 40–42
  in CLN2, 45
  definition, 41
  mechanism of action, 41
  in NCL, 101
  PPT1 interaction, 42
SAPs, *see* Saposins
Seizures
  in Finnish LINCL, 125
  in Northern epilepsy, 126
SEPs, *see* Somatosensory evoked potentials
Sequence homology, CLN3p, 52–53
Signaling, NTFs, failure, 177–178
Single-strand conformation polymorphism, for JNCL mutations, 147
Somatosensory evoked potentials, in Finnish LINCL, 127
Southern hybridization, for *CLN3* deletion, 149
South Hampshire sheep, as human disorder model, 188, 193
Sphingolipid activator proteins, 101
SSCP, *see* Single-strand conformation polymorphism
Storage, *see* Lysosomal storage

T

Testing
  genetic, *see* Genetic testing
  prenatal, for NCL, 152–153
Therapeutics
  NTFs as agents, 173–176, 179
  yeast model for JNCL, 213–214
Tibetan terrier dog, as human disorder model, 193–194
Tissues
  CLN3 expression, 116
  CLN5 expression, 134
  CLN8 expression, 135
  NCL genetic testing, 144
TPP1, *see* Tripeptidyl-peptidase 1
Tripeptidyl-peptidase 1
  in classic late-infanile NCL, 18–19
  CLN2p role, 97–98
  mitochondrial ATP synthase subunit c as substrate, 47–48
  mutations, 47

Subject Index

Turkish late-infantile neuronal ceroid
  lipofuscinosis, 25, 29, 57–58

U

Ultrastructure
  Finnish LINCL, 129–130
  Northern epilepsy, 131–132

V

Vacuolar H^+-ATPase, 53
V-ATPase, see Vacuolar H^+-ATPase

VEP, see Visual evoked potentials
Visual evoked potentials, in Finnish LINCL,
  127

Y

YACs, see Yeast artificial chromosomes
Yeast
  as INCL model, 214
  as JNCL model, 206–209, 213–214
Yeast artificial chromosomes, for CLN3
  mapping, 111–112